影印版说明

MOMENTUM PRESS 出版的 *Plastics Technology Handbook*（2卷）是介绍塑料知识与技术的大型综合性手册，内容涵盖了从高分子基本原理，到塑料的合成、种类、性能、配料、加工、制品，以及模具、二次加工等各个方面。通过阅读、学习本手册，无论是专业人员还是非专业人员，都会很快熟悉和掌握塑料制品的设计和制造方法。可以说一册在手，别无他求。

原版2卷影印时分为11册，第1卷分为：

塑料基础知识 · 塑料性能

塑料制品生产

注射成型

挤压成型

吹塑成型

热成型

发泡成型 · 压延成型

第2卷分为：

涂层 · 浇注成型 · 反应注射成型 · 旋转成型

压缩成型 · 增强塑料 · 其他工艺

模具

辅机与二次加工设备

唐纳德 V · 罗萨多，波士顿大学化学学士学位，美国东北大学 MBA 学位，马萨诸塞大学洛厄尔分校工程塑料和加州大学工商管理博士学位（伯克利）。著有诸多论文及著作，包括《塑料简明百科全书》、《注塑手册（第三版）》以及塑料产品材料和工艺选择手册等。活跃于塑料界几十年，现任著名的 Plasti Source Inc. 公司总裁，并是美国塑料工业协会（SPI）、美国塑料学会（PIA）和 SAMPE（The Society for the Advancement of Material and Process Engineering）的重要成员。

材料科学与工程图书工作室

联系电话 0451-86412421

0451-86414559

邮　箱 yh_bj@aliyun.com

xuyaying81823@gmail.com

zhxh6414559@aliyun.com

影印版

PLASTICS TECHNOLOGY HANDBOOK

塑料技术手册

VOLUME 2

COATING·CASTING·REACTION INJECTION MOLDING·ROTATIONAL MOLDING

涂层·浇注成型·反应注射成型·旋转成型

EDITED BY
DONALD V. ROSATO
MARLENE G. ROSATO
NICK R. SCHOTT

哈爾濱工業大學出版社
HITP
HARBIN INSTITUTE OF TECHNOLOGY PRESS

黑版贸审字08-2014-093号

Donald V.Rosato,Marlene G.Rosato,Nick R.Schott
Plastics Technology Handbook Volume 2
9781606500828

Originally published by Momentum Press, LLC
English reprint rights arranged with Momentum Press, LLC through McGraw-Hill Education (Asia)

图书在版编目（CIP）数据

塑料技术手册. 第2卷. 涂层・浇注成型・反应注射成型・旋转成型 =Plastics technology handbook volume 2 coating・casting・reaction injection molding・rotational molding : 英文 /（美）罗萨多（Rosato, D. V.）等主编. —影印本. —哈尔滨 : 哈尔滨工业大学出版社, 2015.6

ISBN 978-7-5603-5047-9

Ⅰ. ①塑… Ⅱ. ①罗… Ⅲ. ①塑料－技术手册－英文 ②塑料成型－技术手册－英文 Ⅳ. ①TQ320.6-62

中国版本图书馆CIP数据核字（2014）第280094号

责任编辑 许雅莹　张秀华　杨　桦
出版发行 哈尔滨工业大学出版社
社　　址 哈尔滨市南岗区复华四道街10号 邮编 150006
传　　真 0451-86414749
网　　址 http://hitpress.hit.edu.cn
印　　刷 哈尔滨市石桥印务有限公司
开　　本 787mm×960mm　1/16　印张13.25
版　　次 2015年6月第1版　2015年6月第1次印刷
书　　号 ISBN 978-7-5603-5047-9
定　　价 66.00元

（如因印刷质量问题影响阅读，我社负责调换）

PLASTICS TECHNOLOGY HANDBOOK

VOLUME 2

EDITED BY

Donald V. Rosato, PhD, MBA, MS, BS, PE
PlastiSource Inc.
Society of Plastics Engineers
Plastics Pioneers Association
UMASS Lowell Plastics Advisory Board

Marlene G. Rosato, BASc (ChE), P Eng
Gander International Inc.
Canadian Society of Chemical Engineers
Association of Professional Engineers of Ontario
Product Development and Management Association

Nick R. Schott, PhD, MS, BS (ChE), PE
UMASS Lowell Professor of Plastics Engineering Emeritus & Plastics Department Head Retired
Plastics Institute of America
Secretary & Director for Educational and Research Programs

Momentum Press, LLC, New York

Contents

Figures

Tables

ABBREVIATIONS

AA acrylic acid
AAE American Association of Engineers
AAES American Association of Engineering Societies
ABR polyacrylate
ABS acrylontrile-butadiene-styrene
AC alternating current
ACS American Chemical Society
ACTC Advanced Composite Technology Consortium
ad adhesive
ADC allyl diglycol carbonate (also CR-39)
AFCMA Aluminum Foil Container Manufacturers' Association
AFMA American Furniture Manufacturers' Association
AFML Air Force Material Laboratory
AFPA American Forest and Paper Association
AFPR Association of Foam Packaging Recyclers
AGMA American Gear Manufacturers' Association
AIAA American Institute of Aeronautics and Astronauts
AIChE American Institute of Chemical Engineers
AIMCAL Association of Industrial Metallizers, Coaters, and Laminators
AISI American Iron and Steel Institute
AMBA American Mold Builders Association
AMC alkyd molding compound
AN acrylonitrile
ANSI American National Standards Institute
ANTEC Annual Technical Conference (of the Society of the Plastic Engineers)
APC American Plastics Council
APET amorphous polyethylene terephthalate
APF Association of Plastics Fabricators
API American Paper Institute
APME Association of Plastics Manufacturers in Europe
APPR Association of Post-Consumer Plastics Recyclers
AQL acceptable quality level
AR aramid fiber; aspect ratio
ARP advanced reinforced plastic
ASA acrylonitrile-styrene-acrylate
ASCII american standard code for information exchange
ASM American Society for Metals

ASME American Society of Mechanical Engineers
ASNDT American Society for Non-Destructive Testing
ASQC American Society for Quality Control
ASTM American Society for Testing Materials
atm atmosphere
bbl barrel
BFRL Building and Fire Research Laboratory
Bhn Brinell hardness number
BM blow molding
BMC bulk molding compound
BO biaxially oriented
BOPP biaxially oriented polypropylene
BR polybutadiene
Btu British thermal unit
buna polybutadiene
butyl butyl rubber
CA cellulose acetate
CAB cellulose acetate butyrate
$CaCO_3$ calcium carbonate (lime)
CAD computer-aided design
CAE computer-aided engineering
CAM computer-aided manufacturing
CAMPUS computer-aided material preselection by uniform standards
CAN cellulose acetate nitrate
CAP cellulose acetate propionate
CAS Chemical Abstract Service (a division of the American Chemical Society)
CAT computer-aided testing
CBA chemical blowing agent
CCA cellular cellulose acetate
CCV Chrysler composites vehicle
CEM Consorzio Export Mouldex (Italian)
CFA Composites Fabricators Association
CFC chlorofluorocarbon
CFE polychlorotrifluoroethylene
CIM ceramic injection molding; computer integrated manufacturing
CLTE coefficient of linear thermal expansion
CM compression molding
CMA Chemical Manufacturers' Association
CMRA Chemical Marketing Research Association
CN cellulose nitrate (celluloid)
CNC computer numerically controlled
CP Canadian Plastics
CPE chlorinated polyethylene
CPET crystallized polyethylene terephthalate
CPI Canadian Plastics Institute
cpm cycles/minute
CPVC chlorinated polyvinyl chloride
CR chloroprene rubber; compression ratio
CR-39 allyl diglycol carbonate
CRP carbon reinforced plastics
CRT cathode ray tube
CSM chlorosulfonyl polyethylene
CTFE chlorotrifluorethylene
DAP diallyl phthalate
dB decibel
DC direct current
DEHP diethylhexyl phthalate
den denier
DGA differential gravimetric analysis
DINP diisononyl phthalate
DMA dynamic mechanical analysis
DMC dough molding compound
DN *Design News* publication
DOE Design of Experments
DSC differential scanning calorimeter
DSD Duales System Deutschland (German Recycling System)
DSQ German Society for Quality
DTA differential thermal analysis
DTGA differential thermogravimetric analysis
DTMA dynamic thermomechanical analysis
DTUL deflection temperature under load
DV devolatilization
DVR design value resource; dimensional velocity research; Druckverformungsrest (German

compression set); dynamic value research; dynamic velocity ratio
E modulus of elasticity; Young's modulus
EBM extrusion blow molding
E_c modulus, creep (apparent)
EC ethyl cellulose
ECTFE polyethylene-chlorotrifluoroethylene
EDM electrical discharge machining
E/E electronic/electrical
EEC European Economic Community
EI modulus × moment of inertia (equals stiffness)
EMI electromagnetic interference
EO ethylene oxide (also EtO)
EOT ethylene ether polysulfide
EP ethylene-propylene
EPA Environmental Protection Agency
EPDM ethylene-propylene diene monomer
EPM ethylene-propylene fluorinated
EPP expandable polypropylene
EPR ethylene-propylene rubber
EPS expandable polystyrene
E_r modulus, relaxation
E_s modulus, secant
ESC environmental stress cracking
ESCR environmental stress cracking resistance
ESD electrostatic safe discharge
ET ethylene polysulfide
ETFE ethylene terafluoroethylene
ETO ethylene oxide
EU entropy unit; European Union
EUPC European Association of Plastics Converters
EUPE European Union of Packaging and Environment
EUROMAP Eu^ropean Committee of Machine Manufacturers for the Rubber and Plastics Industries (Zurich, Switzerland)
EVA ethylene-vinyl acetate
E/VAC ethylene/vinyl acetate copolymer
EVAL ethylene-vinyl alcohol copolymer (tradename for EVOH)
EVE ethylene-vinyl ether
EVOH ethylene-vinyl alcohol copolymer (or EVAL)
EX extrusion
F coefficient of friction; Farad; force
FALLO follow all opportunities
FDA Food and Drug Administration
FEA finite element analysis
FEP fluorinated ethylene-propylene
FFS form, fill, and seal
FLC fuzzy logic control
FMCT fusible metal core technology
FPC flexible printed circuit
fpm feet per minute
FRCA Fire Retardant Chemicals Association
FRP fiber reinforced plastic
FRTP fiber reinforced thermoplastic
FRTS fiber reinforced thermoset
FS fluorosilicone
FTIR Fourier transformation infrared
FV frictional force × velocity
G gravity; shear modulus (modulus of rigidity); torsional modulus
GAIM gas-assisted injection molding
gal gallon
GB gigabyte (billion bytes)
GD&T geometric dimensioning and tolerancing
GDP gross domestic product
GFRP glass fiber reinforced plastic
GMP good manufacturing practice
GNP gross national product
GP general purpose
GPa giga-Pascal
GPC gel permeation chromatography
gpd grams per denier
gpm gallons per minute
GPPS general purpose polystyrene
GRP glass reinforced plastic
GR-S polybutadiene-styrene
GSC gas solid chromatography

H hysteresis; hydrogen
HA hydroxyapatite
HAF high-abrasion furnace
HB Brinell hardness number
HCFC hydrochlorofluorocarbon
HCl hydrogen chloride
HDPE high-density polyethylene (also PE-HD)
HDT heat deflection temperature
HIPS high-impact polystyrene
HMC high-strength molding compound
HMW-HDPE high molecular weight–high density polyethylene
H-P Hagen-Poiseuille
HPLC high-pressure liquid chromatography
HPM hot pressure molding
HTS high-temperature superconductor
Hz Hertz (cycles)
I integral; moment of inertia
IB isobutylene
IBC internal bubble cooling
IBM injection blow molding; International Business Machines
IC *Industrial Computing* publication
ICM injection-compression molding
ID internal diameter
IEC International Electrochemical Commission
IEEE Institute of Electrical and Electronics Engineers
IGA isothermal gravimetric analysis
IGC inverse gas chromatography
IIE Institute of Industrial Engineers
IM injection molding
IMM injection molding machine
IMPS impact polystyrene
I/O input/output
ipm inch per minute
ips inch per second
IR synthetic polyisoprene (synthetic natural rubber)
ISA Instrumentation, Systems, and Automation
ISO International Standardization Organization or International Organization for Standardization
IT information technology
IUPAC International Union of Pure and Applied Chemistry
IV intrinsic viscosity
IVD in vitro diagnostic
J joule
JIS Japanese Industrial Standard
JIT just-in-time
JIT just-in-tolerance
J_p polar moment of inertia
JSR Japanese SBR
JSW Japan Steel Works
JUSE Japanese Union of Science and Engineering
JWTE Japan Weathering Test Center
K bulk modulus of elasticity; coefficient of thermal conductivity; Kelvin; Kunststoffe (plastic in German)
kb kilobyte (1000 bytes)
kc kilocycle
kg kilogram
KISS keep it short and simple
Km kilometer
kPa kilo-Pascal
ksi thousand pounds per square inch (psi $\times 10^3$)
lbf pound-force
LC liquid chromatography
LCP liquid crystal polymer
L/D length-to-diameter (ratio)
LDPE low-density polyethylene (PE-LD)
LIM liquid impingement molding; liquid injection molding
LLDPE linear low-density polyethylene (also PE-LLD)
LMDPE linear medium density polyethylene
LOX liquid oxygen
LPM low-pressure molding
m matrix; metallocene (catalyst); meter

mμ micromillimeter; millicron; 0.000001 mm
μm micrometer
MA maleic anhydride
MAD mean absolute deviation; molding area diagram
Mb bending moment
MBTS benzothiazyl disulfide
MD machine direction; mean deviation
MD&DI Medical Device and Diagnostic Industry
MDI methane diisocyanate
MDPE medium density polyethylene
Me metallocene catalyst
MF melamine formaldehyde
MFI melt flow index
mHDPE metallocene high-density polyethylene
MI melt index
MIM metal powder injection molding
MIPS medium impact polystyrene
MIT Massachusetts Institute of Technology
mLLDPE metallocene catalyst linear low-density polyethylene
MMP multimaterial molding or multimaterial multiprocess
MPa mega-Pascal
MRPMA Malaysian Rubber Products Manufacturers' Association
Msi million pounds per square inch (psi $\times 10^6$)
MSW municipal solid waste
MVD molding volume diagram
MVT moisture vapor transmission
MW molecular weight
MWD molecular weight distribution
MWR molding with rotation
N Newton (force)
NACE National Association of Corrosion Engineers
NACO National Association of CAD/CAM Operation
NAGS North America Geosynthetics Society
NASA National Aeronautics Space Administration
NBR butadiene acrylontrile
NBS National Bureau of Standards (since 1980 renamed the National Institute Standards and Technology or NIST)
NC numerical control
NCP National Certification in Plastics
NDE nondestructive evaluation
NDI nondestructive inspection
NDT nondestructive testing
NEAT nothing else added to it
NEMA National Electrical Manufacturers' Association
NEN Dutch standard
NFPA National Fire Protection Association
NISO National Information Standards Organization
NIST National Institute of Standards and Technology
nm nanometer
NOS not otherwise specified
NPCM National Plastics Center and Museum
NPE National Plastics Exhibition
NPFC National Publications and Forms Center (US government)
NR natural rubber (polyisoprene)
NSC National Safety Council
NTMA National Tool and Machining Association
NWPCA National Wooden Pallet and Container Association
OD outside diameter
OEM original equipment manufacturer
OPET oriented polyethylene terephthalate
OPS oriented polystyrene
OSHA Occupational Safety and Health Administration
P load; poise; pressure
Pa Pascal
PA polyamide (nylon)
PAI polyamide-imide
PAN polyacrylonitrile

PB polybutylene
PBA physical blowing agent
PBNA phenyl-β-naphthylamine
PBT polybutylene terephthalate
PC permeability coefficient; personal computer; plastic composite; plastic compounding; plastic-concrete; polycarbonate; printed circuit; process control; programmable circuit; programmable controller
PCB printed circuit board
pcf pounds per cubic foot
PCFC polychlorofluorocarbon
PDFM Plastics Distributors and Fabricators Magazine
PE plastic engineer; polyethylene (UK polythene); professional engineer
PEEK polyetheretherketone
PEI polyetherimide
PEK polyetherketone
PEN polyethylene naphthalate
PES polyether sulfone
PET polyethylene terephthalate
PETG polyethylene terephthalate glycol
PEX polyethylene crosslinked pipe
PF phenol formaldehyde
PFA perfluoroalkoxy (copolymer of tetrafluoroethylene and perfluorovinylethers)
PFBA polyperfluorobutyl acrylate
phr parts per hundred of rubber
PI polyimide
PIA Plastics Institute of America
PID proportional-integral-differential
PIM powder injection molding
PLASTEC Plastics Technical Evaluation Center (US Army)
PLC programmable logic controller
PMMA Plastics Molders and Manufacturers' Association (of SME); polymethyl methacrylate (acrylic)
PMMI Packaging Machinery Manufacturers' Institute
PO polyolefin
POE polyolefin elastomer
POM polyoxymethylene or polyacetal (acetal)
PP polypropylene
PPA polyphthalamide
ppb parts per billion
PPC polypropylene chlorinated
PPE polyphenylene ether
pph parts per hundred
ppm parts per million
PPO polyphenylene oxide
PPS polyphenylene sulfide
PPSF polyphenylsulfone
PPSU polyphenylene sulphone
PS polystyrene
PSB polystyrene butadiene rubber (GR-S, SBR)
PS-F polystyrene-foam
psf pounds per square foot
PSF polysulphone
psi pounds per square inch
psia pounds per square inch, absolute
psid pounds per square inch, differential
psig pounds per square inch, gauge (above atmospheric pressure)
PSU polysulfone
PTFE polytetrafluoroethylene (or TFE)
PUR polyurethane (also PU, UP)
P-V pressure-volume (also PV)
PVA polyvinyl alcohol
PVAC polyvinyl acetate
PVB polyvinyl butyral
PVC polyvinyl chloride
PVD physical vapor deposition
PVDA polyvinylidene acetate
PVdC polyvinylidene chloride
PVDF polyvinylidene fluoride
PVF polyvinyl fluoride
PVP polyvinyl pyrrolidone

PVT pressure-volume-temperature (also P-V-T or pvT)
PW *Plastics World* magazine
QA quality assurance
QC quality control
QMC quick mold change
QPL qualified products list
QSR quality system regulation
R Reynolds number; Rockwell (hardness)
rad Quantity of ionizing radiation that results in the absorption of 100 ergs of energy per gram of irradiated material.
radome radar dome
RAPRA Rubber and Plastics Research Association
RC Rockwell C (R_c)
RFI radio frequency interference
RH relative humidity
RIM reaction injection molding
RM rotational molding
RMA Rubber Manufacturers' Association
RMS root mean square
ROI return on investment
RP rapid prototyping; reinforced plastic
RPA Rapid Prototyping Association (of SME)
rpm revolutions per minute
RRIM reinforced reaction injection molding
RT rapid tooling; room temperature
RTM resin transfer molding
RTP reinforced thermoplastic
RTS reinforced thermoset
RTV room temperature vulcanization
RV recreational vehicle
Rx radiation curing
SAE Society of Automotive Engineers
SAMPE Society for the Advancement of Material and Process Engineering
SAN styrene acrylonitrile
SBR styrene-butadiene rubber
SCT soluble core technology
SDM standard deviation measurement
SES Standards Engineering Society
SF safety factor; short fiber; structural foam
s.g. specific gravity
SI International System of Units
SIC Standard Industrial Classification
SMC sheet molding compound
SMCAA Sheet Molding Compound Automotive Alliance
SME Society of Manufacturing Engineers
S-N stress-number of cycles
SN synthetic natural rubber
SNMP simple network management protocol
SPC statistical process control
SPE Society of the Plastics Engineers
SPI Society of the Plastics Industry
sPS syndiotactic polystyrene
sp. vol. specific volume
SRI Standards Research Institute (ASTM)
S-S stress-strain
STP Special Technical Publication (ASTM); standard temperature and pressure
t thickness
T temperature; time; torque (or T_t)
TAC triallylcyanurate
T/C thermocouple
TCM technical cost modeling
TD transverse direction
TDI toluene diisocyanate
TF thermoforming
TFS thermoform-fill-seal
T_g glass transition temperature
TGA thermogravimetric analysis
TGI thermogravimetric index
TIR tooling indicator runout
T-LCP thermotropic liquid crystal polymer
TMA thermomechanical analysis; Tooling and Manufacturing Association (formerly TDI); Toy Manufacturers of America
torr mm mercury (mmHg); unit of pressure equal to 1/760th of an atmosphere

TP thermoplastic
TPE thermoplastic elastomer
TPO thermoplastic olefin
TPU thermoplastic polyurethane
TPV thermoplastic vulcanizate
T_s tensile strength; thermoset
TS twin screw
TSC thermal stress cracking
TSE thermoset elastomer
TX thixotropic
TXM thixotropic metal slurry molding
UA urea, unsaturated
UD unidirectional
UF urea formaldehyde
UHMWPE ultra-high molecular weight polyethylene (also PE-UHMW)
UL Underwriters Laboratories
UP unsaturated polyester (also TS polyester)
UPVC unplasticized polyvinyl chloride
UR urethane (also PUR, PU)
URP unreinforced plastic
UV ultraviolet
UVCA ultra-violet-light-curable-cyanoacrylate
V vacuum; velocity; volt
VA value analysis
VCM vinyl chloride monomer
VLDPE very low-density polyethylene
VOC volatile organic compound
vol% percentage by volume
w width
W watt
W/D weight-to-displacement volume (boat hull)
WIT water-assist injection molding technology
WMMA Wood Machinery Manufacturers of America
WP&RT World Plastics and Rubber Technology magazine
WPC wood-plastic composite
wt% percentage by weight
WVT water vapor transmission
XL cross-linked
XLPE cross-linked polyethylene
XPS expandable polystyrene
YPE yield point elongation
Z-twist twisting fiber direction

Acknowledgments

Undertaking the development through to the completion of the *Plastics Technology Handbook* required the assistance of key individuals and groups. The indispensable guidance and professionalism of our publisher, Joel Stein, and his team at Momentum Press was critical throughout this enormous project. The coeditors, Nick R. Schott, Professor Emeritus of the University of Massachusetts Lowell Plastics Engineering Department, and Marlene G. Rosato, President of Gander International Inc., were instrumental to the data, information, and analysis coordination of the eighteen chapters of the handbook. A special thank you is graciously extended to Napoleao Neto of Alphagraphics for the organization and layout of the numerous figure and table graphics central to the core handbook theme. Finally, a great debt is owed to the extensive technology resources of the Plastics Institute of America at the University of Massachusetts Lowell and its Executive Director, Professor Aldo M. Crugnola.

Dr. Donald V. Rosato, Coeditor and President, PlastiSource, Inc.

Preface

This book, as a two-volume set, offers a simplified, practical, and innovative approach to understanding the design and manufacture of products in the world of plastics. Its unique review will expand and enhance your knowledge of plastic technology by defining and focusing on past, current, and future technical trends. Plastics behavior is presented to enhance one's capability when fabricating products to meet performance requirements, reduce costs, and generally be profitable. Important aspects are also presented to help the reader gain understanding of the advantages of different materials and product shapes. The information provided is concise and comprehensive.

Prepared with the plastics technologist in mind, this book will be useful to many others. The practical and scientific information contained in this book is of value to both the novice, including trainees and students, and the most experienced fabricators, designers, and engineering personnel wishing to extend their knowledge and capability in plastics manufacturing including related parameters that influence the behavior and characteristics of plastics. The toolmaker (who makes molds, dies, etc.), fabricator, designer, plant manager, material supplier, equipment supplier, testing and quality control personnel, cost estimator, accountant, sales and marketing personnel, new venture type, buyer, vendor, educator/trainer, workshop leader, librarian, industry information provider, lawyer, and consultant can all benefit from this book. The intent is to provide a review of the many aspects of plastics that range from the elementary to the practical to the advanced and more theoretical approaches. People with different interests can focus on and interrelate across subjects in order to expand their knowledge within the world of plastics.

Over 20000 subjects covering useful pertinent information are reviewed in different chapters contained in the two volumes of this book, as summarized in the expanded table of contents and index. Subjects include reviews on materials, processes, product designs, and so on. From a pragmatic standpoint, any theoretical aspect that is presented has been prepared so that the practical person will understand it and put it to use. The theorist in turn will gain an insight into the practical limitations that exist in plastics as they exist in other materials such as steel, wood, and so on. There is no material that is "perfect." The two volumes of this book together contain 1800-plus figures and 1400-plus tables providing extensive details to supplement the different subjects.

In working with any material (plastics, metal, wood, etc.), it is important to know its behavior in order to maximize product performance relative to cost and efficiency. Examples of different plastic materials and associated products are reviewed with their behavior patterns. Applications span toys, medical devices, cars, boats, underwater devices, containers, springs, pipes, buildings, aircraft, and spacecraft. The reader's product to be designed or fabricated, or both, can be related directly or indirectly to products reviewed in this book. Important are behaviors associated with and interrelated with the many different plastics materials (thermoplastics [TPs], thermosets [TSs], elastomers, reinforced plastics) and the many fabricating processes (extrusion, injection molding, blow molding, forming, foaming, reaction injection molding, and rotational molding). They are presented so that the technical or nontechnical reader can readily understand the interrelationships of materials to processes.

This book has been prepared with the awareness that its usefulness will depend on its simplicity and its ability to provide essential information. An endless amount of data exists worldwide for the many plastic materials, which total about 35000 different types. Unfortunately, as with other materials, a single plastic material that will meet all performance requirements does not exist. However, more so than with any other materials, there is a plastic that can be used to meet practically any product requirement. Examples are provided of different plastic products relative to critical factors ranging from meeting performance requirements in different environments to reducing costs and targeting for zero defects. These reviews span products that are small to large and of shapes that are simple to complex. The data included provide examples that span what is commercially available. For instance, static physical properties (tensile, flexural, etc.), dynamic physical properties (creep, fatigue, impact, etc.), chemical properties, and so on, can range from near zero to extremely high values, with some having the highest of any material. These plastics can be applied in different environments ranging from below and on the earth's surface to outer space.

Pitfalls to be avoided are reviewed in this book. When qualified people recognize the potential problems, these problems can be designed around or eliminated so that they do not affect the product's performance. In this way, costly pitfalls that result in poor product performance or failure can be reduced or eliminated. Potential problems or failures are reviewed, with solutions also presented. This failure-and-solution review will enhance the intuitive skills of people new to plastics as well as those who are already working in plastics. Plastic materials have been produced worldwide over many years for use in the design and fabrication of all kinds of plastic products. To profitably and successfully meet high-quality, consistency, and long-life standards, all that is needed is to understand the behavior of plastics and to apply these behaviors properly.

Patents or trademarks may cover certain of the materials, products, or processes presented. They are discussed for information purposes only and no authorization to use these patents or trademarks is given or implied. Likewise, the use of general descriptive names, proprietary names, trade names, commercial designations, and so on does not in any way imply that they may be used freely. While the information presented represents useful information that can be studied or analyzed and is believed to be true and accurate, neither the authors, contributors, reviewers, nor the publisher can accept any legal responsibility for any errors, omissions, inaccuracies, or other factors. Information is provided without warranty of any kind. No representation as to accuracy, usability, or results should be inferred.

Preparation for this book drew on information from participating industry personnel, global industry and trade associations, and the authors' worldwide personal, industrial, and teaching experiences.

DON & MARLENE ROSATO AND NICK SCHOTT, 2011

About the Authors

Dr. Donald V. Rosato, president of PlastiSource Inc., a prototype manufacturing, technology development, and marketing advisory firm in Massachusetts, United States, is internationally recognized as a leader in plastics technology, business, and marketing. He has extensive technical, marketing, and plastics industry business experience ranging from laboratory testing to production to marketing, having worked for Northrop Grumman, Owens-Illinois, DuPont/Conoco, Hoechst Celanese/Ticona, and Borg Warner/G.E. Plastics. He has developed numerous polymer-related patents and is a participating member of many trade and industry groups. Relying on his unrivaled knowledge of the industry and high-level international contacts, Dr. Rosato is also uniquely positioned to provide an expert, inside view of a range of advanced plastics materials, processes, and applications through a series of seminars and webinars. Among his many accolades, Dr. Rosato has been named Engineer of the Year by the Society of Plastics Engineers. Dr. Rosato has written extensively, authoring or editing numerous papers, including articles published in the *Encyclopedia of Polymer Science and Engineering*, and major books, including the *Concise Encyclopedia of Plastics*, *Injection Molding Handbook 3rd ed.*, *Plastic Product Material and Process Selection Handbook*, *Designing with Plastics and Advanced Composites*, and *Plastics Institute of America Plastics Engineering, Manufacturing, and Data Handbook*. Dr. Rosato holds a BS in chemistry from Boston College, an MBA from Northeastern University, an MS in plastics engineering from the University of Massachusetts Lowell, and a PhD in business administration from the University of California, Berkeley.

Marlene G. Rosato, with stints in France, China, and South Korea, has comprehensive international plastics and elastomer business experience in technical support, plant start-up and troubleshooting, manufacturing and engineering management, and business development and strategic planning with Bayer/Polysar and DuPont. She also does extensive international technical, manufacturing, and management consulting as president of Gander International Inc. She also has an extensive writing background authoring or editing numerous papers and major books, including the *Concise Encyclopedia of Plastics*, *Injection Molding Handbook 3rd ed.*, and the *Plastics Institute of America Plastics Engineering, Manufacturing and Data Handbook*. A senior member of the Canadian Society of Chemical Engineering and the Association of Professional Engineers of Canada, Ms. Rosato is a licensed professional engineer of Ontario, Canada. She received a Bachelor of Applied Science in chemical engineering from the University of British Columbia with continuing education at McGill University in Quebec, Queens University and the University of Western Ontario, both in Ontario, and also has extensive executive management training.

Emeritus Professor Nick Schott, a long-time member of the world-renowned University of Massachusetts Lowell Plastics Engineering Department faculty, served as its department head for a quarter of a century. Additionally, he founded the Institute for Plastics Innovation, a research consortium affiliated with the university that conducts research related to plastics manufacturing, with a current emphasis on bioplastics, and served as its director from 1989 to 1994. Dr. Schott has received numerous plastics industry accolades from the SPE, SPI, PPA, PIA, as well as other global industry associations and is renowned for the depth of his plastics technology experience, particularly in processing-related areas. Moreover, he is a quite prolific and requested industry presenter, author, patent holder, and product/process developer. In addition, he has extensive and continuing academic responsibilities at the undergraduate to postdoctoral levels. Among America's internationally recognized plastics professors, Dr. Nick R. Schott most certainly heads everyone's list not only within the 2500 plus global UMASS Lowell Plastics Engineering alumni family, which he has helped grow, but also in broad global plastics and industrial circles. Professor Schott holds a BS in chemical engineering from UC Berkeley, and an MS and PhD from the University of Arizona.

CHAPTER 10

COATING

OVERVIEW

Different resin (also called polymer and plastic) coating systems have widespread industrial and commercial applications. They can be applied by direct contact of a liquid coating with the substrate to deposition using an atomization process. Direct methods include brushing, roller coating, dipping, flow coating, and electrodeposition. Deposition methods include conventional spray, airless spray, hot spray, and electrostatic spray. Extrusion coating is one of the principal methods (chapter 5). Coating via calenders is another important method (chapter 9). Coatings are applied in molds during injection molding (chapter 4). There is film coating applied during thermoforming (chapter 7; 477). Other fabricating processes incorporate coatings.

Coating resins are used for coating materials in practically all the markets that include electric/electronic, packaging, building, household and industrial appliances, transportation, marine, medical, and clothing (chapter 20). Continual consumer demands for more attractive and styled packages have caused plastic material suppliers to develop new coatings with high decorative and visual appeal. Selection of the plastic to be used usually depends on decorative and environmental requirements. Coated containers include beer cans, liquid-containing tanks, and electronics packages.

Resin coatings are used extensively for corrosion protection of metals in different environments such as inside and outside of buildings, chemical plants, marine products, and so on. Residual stresses can be present in these coatings. For example, solvent loss and, in the case of thermosets (TSs), the curing process, causes shrinkage of the coating. When it is applied to a stiff substrate, the shrinkage in the plane of the coating is resisted and biaxial tensile residual stresses form. If application of the coating is made at a temperature different from the subsequent service temperature, there will be further residual stresses that result from differential thermal expansion of the coating and substrate (chapter 21).

Resins continue to be the backbone in the coating industry because almost all coatings are composed of resin materials. The most widely used are based on polyethylenes, polypropylenes, vinyls, alkyds, acrylics, urea-melamine, styrenes, epoxies, phenolics, fluoroplastics, and silicones (chapter 2). The resins are used alone or are cross-blended with other resins. Table 10.1 lists different coatings that include those in solvent systems or those with certain resins, such as vinyl in different forms of organic media dispersions. These higher solids content dispersions can be in nonvolatile (plastisol) suspension or in volatile (organosol) suspension (chapter 16). Table 10.2 reviews coating compounds with applications that are mainly used in Europe.

The alkyds are used mainly (but extensively) in coatings. Their ease of application and low cost makes them useful. Epoxy systems continue to find more applications because they have inherently desirable characteristics such as the ease with which they adhere to a substrate. Fluorocarbons can be vacuum deposited on various metals and plastics containers, which provides the expected environmental resistances, such as to water and salt spray. The polyamides are used to protect metal containers from weathering and chemical corrosion. The silicones are considered for use when heat resistance is part of the coating requirement. Urethane coatings are generally baked so to provide maximum protection in such applications as electrical or outdoor-use packages. Properties of the different plastics are reviewed in chapter 2.

The vinyls appear to be in a class of their own, because they can be applied by many different techniques to metal and other parts before fabrication into various shapes. They are tough, flexible, low in cost, and resistant to normal environments. They also provide good adhesion.

Fuel-resistant coatings are used to help the handling of gasoline and fuel oil in plastic or steel tanks. Coating systems employed in the past were only partially successful in protecting the steel interior of the tanks. They often cracked, peeled, or softened and thus exposed the steel to corrosion. Excellent coatings have now been developed and used successfully.

Growth has been steady and reliable, so that rational and economic coating production is no longer regarded, as was the case until relatively recently, as an art or craft based solely on empirical results. For example, coatings are being applied to plastic and metal containers to provide improvements in appearance, resistance to environmental breakdown, and easy handling.

Resins are employed in the manufacture of a large number of coating compounds that are used to cover the surfaces of many materials from paper to metal to concrete. Many plastics are used as coating materials, including paints, varnishes, enamels, and materials of various resin-coating compositions that are applied to fabrics, paper, plastic, cardboard, leather, metal, and wood. As has been noted, there are a number of methods of applying plastic coatings, including the major processes such as extrusion and injection molding. When they are in a liquid or latex form, such as in paint or adhesives, they may be applied by brushing, spraying, dipping, and so on. In coating operations, the base material, such as paper, is run through a machine that melts solid plastic granules and spreads them evenly across the surface of the base material. As the hot plastic cools, it becomes bonded, like an adhesive, to its base.

The products of the coating industry are essential for the protection and decoration of the majority of manufactured goods and architectural or industrial structures that characterize our

Type	Characteristics
Automotive Paints:	
Urethane Enamel	The best of the car finishes, lasts over 10 years, has the best look, and is the most expensive. Paint jobs can run over $1000 and paint cost alone ranges from $50 to $100 per gallon.
"Clear Coat"	This is the top coat of a two part paint. The Clear Coat is applied over a base coat of acrylic enamel or acrylic lacquer and produces a beautiful "wet look" finish just like a factory paint job. This type of finish is very difficult to apply and should be done by an expert. Has a life of 8 to 10 years and costs between $400 & $600.
Acrylic Lacquer	Mid range auto paint, very fast drying, much higher gloss and better durability than the alkyd enamels. Must be machine polished after drying so it is more expensive than the Acrylic Enamel paints. Acrylic lacquer must not be painted over acrylic enamel.Expect to pay $300 to $500 for this paint job. Life span is 5 to 7 years.
Acrylic Enamel	Mid range auto paint, very slow drying, much higher gloss and better durability than the alkyd enamels and acrylic lacquer. Acrylic enamel should not be painted over acrylic lacquer. Usually requires a heat booth to aid drying. Expect to pay $200 to $300 for a paint job.
Alkyd Enamel	Cheap paint with low durability (will some–times loose its gloss in less than 2 months). Paint life will only be 1 to 3 years. The paint job will probably only cost $100 to $200 and is commonly referred to as the "baked enamel" job since the vehicle is baked at 150°F in a heat booth to set the paint.
Lacquer	Lacquer is a fast–drying, high gloss varnish used by most furniture manufacturers as the top–coat finish. It is very hard, dries crystal clear and is highly resistant to alcohols, water, heat, and mild acids. Although the original lacquers came from insects and the sumac tree, almost all produced today are synthetic and are mixed with some combination of resins (better adhesion), nitrocellulose, linseed or castor oil (improves flexibility), vinyls, acrylics or synthetic polymers. The main problem with lacquers is that they dry so fast that it is sometimes difficult to get a good finish. Use a spray gun if possible. Multiple coats are usually necessary

Table 10.1 Examples of different coating materials

House Paints	There are basically 5 groups of house paints. Oil Base, Alkyd, Emulsion, Water Thinned, and Catalytic. Each of these classes is subdivided into Exterior and Interior.
Oil Base	Interior and Exterior, oil vehicle, thinned by solvents such as turpentine and mineral spirits, very slow drying, strong smell. Mainly used as Exterior paint. Use in well ventilated area. Good adhesion to chalky surfaces.
Alkyd	Synthetic oil vehicle of a resin known as Alkyd. Interior and Exterior enamels, easy to apply, fast drying, odorless, and produce a tough coating. Easy cleanup and thinning with mineral spirits. Excellent interior paint, not resistant to chemicals, solvents, or corrosives.
Emulsion	Water based paint mixture. Latex paints fall into this category and the most common are acrylic and vinyl (PVA). Available as interior and exterior, and as flat, gloss, and semigloss enamels. Very quick drying (sometimes less than 1 hour) but do not wash for 2 to 4 weeks, paints over damp surfaces, odorless, alkalis resistant, doesn't usually blister and peel. Excellent cover and blending characteristics, but poor adhesion to chalky surfaces, easy cleanup. Use special latex primers for painting bare wood. Paint at temperatures above 45°F. By far the most popular paint today.
Water–Thinned	Generally used to describe non–emulsion paint such as calcimine and casein and white wash. These paints are used primarily on masonry surfaces. The most common water thinned paint being Portland Cement Paints.
Catalytic	This class of paints cures by a chemical heat process, not by evaporation of a solvent or water as in the other paints. Catalytic paints are usually two part paints which means that you have to mix two parts to start the curing process. Included in this class are the epoxy and polyurethane resins. They are extremely tough and durable and are highly resistant to water, wear, acids, solvents, abrasion, salt water, and chemicals. Drying times are very fast (several hours). Good ventilation is necessary when working with these paints. Catalytic paints can not be applied over other paint types. Follow the manufacturers instructions very closely, these are not easy paints to use.

Table 10.1 Examples of different coating materials *(continued)*

Exterior Paint	These paints are designed to have long life spans, good adhesion and resistance to moisture, ultraviolet light, mildew, and sulfide and acid fumes. This class also includes the varnish and stain groups described later. Never use interior paints in place of exterior paints, they will not hold up under the weather.
Interior Paint	Interior paints are designed to maximize the hiding ability of the paint with only 1 or 2 coats. Flat paints contain more pigment than high sheen paints but are less durable. Good interior paints can be touched up easily without major changes in the sheen or color.
Varnish	Varnish is a solution of a hard resin, a drying oil, metallics for driers, and solvent. There are two types, natural and synthetic. Natural types are slow drying (24 to 48 hrs) and are subclassed as "long oil" (meaning high oil content) and "short oil" (meaning less oil content). Naturals are tough and used mainly for exterior and marine applications. Synthetics contain resins such as alkyd (the most common), polyurethane, vinyl, and phenolic and are more durable and faster drying than naturals. Apply with natural bristle brushes; apply 3 to 4 coats total, let dry between coats and sand with 240 grit sand paper. Varnishes are usually transparent and are excellent sealers.
Shellac	Shellac is one of the oldest wood finishes. It is made from a mixture of the dry resinous secretions of the lac bug (SE Asia) and alcohol. Once mixed, shellac has a very short shelf life, so store it in flake form. Shellac is mixed in what is called a "cut". A "3 pound cut" is 3 pounds of shellac in 1 gallon of alcohol. Initial coats are typically 1 or 2 pound cuts. Shellac is applied with a brush and the better finishes use 6 to 8 coats. Each coat should be sanded with 220 to 240 grit sand paper after it has dried (1 to 2 hours). The final coat is typically rubbed out with a fine 3/0 steel wool.
Primers	Primers are paints intended to produce a good foundation for the overlying coats of paint. Exterior wood primers penetrate deeply into the surface, adhere tightly to the surface, and seal off the wood. Primers typically have an abundance of pigment to allow sanding if necessary. Metal primers are specifically designed to adhere to the metal and stop any oxidation (rusting). Automotive primers usually have a lot of resin included also.

Table 10.1 Examples of different coating materials *(continued)*

Product groups	Compounds	Application
Silicate paints/products, emulsion paints DIN 18363 (M-SK01)	potash-waterglass (binder), inorganic/organic pigments, mineral fillers, synthetic resins (dispersions, stabilizers), water (dispersing agent)	wall paints, outside house paints
Silicate paints/products, (M-SK02)	potassium silicate (binder, fixing solution), inorganic pigments, adjuvants	water- and weatherproof painting, interior coating (resistant to chemicals)
Emulsion paints (M-DF02)	synthetic resins (dispersions), mineral fillers, inorganic/organic pigments, water (dispersing agent), additives (film-forming agents < 3 %: e.g., glycol-ethers, esters, glycols, hydrocarbons), formaldehyde < 0,1 %	outside house paints, interior coating (wall paints)
Emulsion paints, solvent-free (M-DF01)	synthetic resins (dispersions), mineral fillers, inorganic/organic pigments, water (dispersing agent), additives (film-forming agents: no solvents), formaldehyde < 0,1 %	interior coating (wall paint)
Emulsion paints, outside house paint, water dilutable	emulsions of plastic materials (acrylate, vinyl), pigments, water (dispersing agent), organic solvents: < 0,1-4 %, (glycols, glycol-ethers, mineral varnish)	exterior coating
Emulsion paints, varnishes (M-LW01)	synthetic resins, alkyd resins, copolymerizates, polyurethane resins, inorganic/organic pigments, mineral fillers, additives, water (dispersing agents), organic solvents: 5-10 %, glycols, glycol-ethers, esters, mineral varnish	various applications
Aldehyde resin varnishes, aromatic compounds (M-LL03)	aldehyde resins (binders), inorganic/organic pigments, fillers, 30-55 % solvents: mineral varnish (mixture of hydrocarbons), other solvents (< 10 % esters, ethers, alcohols)	covering varnishes, primers
Aldehyde resin varnishes, low levels of aromatic compounds (M-LL02)	aldehyde resins (binders), inorganic/organic pigments, fillers, 30-55 % solvents: mineral varnish (mixture of hydrocarbons), other solvents (< 10 % esters, ethers, alcohols)	covering varnishes, primers
Aldehyde resin no aromatic substances (M-LL01)	aldehyde resins (binders), inorganic/organic pigments, fillers, 30-55 % solvents: mineral varnish (mixture of hydrocarbons), other solvents (< 10 % esters, ethers, alcohols)	covering varnishes, primers

Table 10.2 Important coating compounds and applications

complex material civilization. The protective function includes resistance to air, water, organic liquids, and aggressive chemicals such as acids and alkalis, together with improved superficial mechanical properties such as greater hardness and abrasion resistance. The decorative effect may be obtained through color, gloss, or texture or combinations of these properties.

In the case of many surfaces such as walls or floors, or objects such as interior fittings, furniture and other articles, the surface coating can fulfill hygienic requirements. The surface should not be prone to collect dirt, bacteria, and other impurities. It should be easy to clean with common cleaning agents. In certain cases special qualities are required of the surface coating. For example, special

Product groups	Compounds	Application
Polymer resin paints, high levels of aromatic compounds (M-PL03)	copolymers, inorganic/organic pigments, fillers, 35-50 % solvents: mineral varnish (mixtures of hydrocarbons), other solvents (< 10 %, esters, ethers, alcohols)	outside paints (mineral background)
Polymer resin paints, low levels of aromatic compounds (M-PL02)	copolymers, inorganic/organic pigments, fillers, 35-50 % solvents: mineral varnish (mixtures of hydrocarbons), other solvents (< 10 %, esters, ethers, alcohols)	outside paints (mineral background)
Polymer resin paints, no aromatic compounds (M-PL01)	copolymers, inorganic/organic pigments, fillers, 35-50 % solvents: mineral varnish (mixtures of hydrocarbons), other solvents (< 10 %, esters, ethers, alcohols)	outside paints (mineral background)
Polymer resin paints, dilutable with solvents (M-PL04)	copolymers, inorganic/organic pigments, fillers, 35-50 % solvents: mineral varnish (mixtures of hydrocarbons), other solvents (< 10 %, esters, ethers, alcohols)	outside paints (mineral background)
Natural resin paints	natural resins (e.g., shellac) or chemical modified natural resins (e.g., colophonium derivates), additions (e.g., methyl cellulose, natural latex, casein), inorganic, organic pigments (mainly natural origin), mineral fillers, additives (organic solvents: alcohols, terpenes, oil of turpentine, limonenes), essential oils (eucalyptus oil, oil of rosemary, oil of bergamot)	various applications
Natural resin paints, solvent-free	natural resins (e.g., shellac) or chemical modified natural resins (e.g., colophonium derivates), additions (e.g., methyl cellulose, natural latex, casein), inorganic, organic pigments (mainly natural origin), mineral fillers, additives (see above): < 1 %	various applications
Oil paints, terpene products (M-LL04)	oils (linseed oil, wood oil, soya oil), natural resins and modified natural resins, mineral pigments, wetting agent, flow improver, solvents: oils of turpentine, isoaliphatics, terpenes (citrus, orange)	covering varnishes, primers
Oil paints, terpene-free (M-LL05)	oils (linseed oil, wood oil, soya oil), natural resins and modified natural resins, mineral pigments, wetting agents, flow improvers, solvents: isoaliphatics (dearomatized)	covering varnishes, primers
Oil paints solvent-free No. 665	standard oils, calcium carbonate, pigments, siccatives, lemon oil water-soluble	exterior coating (paneling)
Clear lacquers/glazing composition (wood), low levels of aromatic compounds (M-KH03)	alkyd resins, nitro resins, polymer resins, pigments, fillers, 40-50 % solvents: mineral varnish (mixtures of hydrocarbons), other solvents (< 10 %, esters, ethers, alcohols)	interior coating (clear lacquers), exterior and interior coatings (glazing compositions, wood)
Lead chromate enamels, aromatic compounds	synthetic resins (e.g., aldehyde resins, PVC-polymerizates) inorganic/ organic pigments (lead chromate up to 20 %), fillers, 30-55 % solvents (mixtures of mineral varnish, glycol-ethers, aromatic compounds)	priming coat (steel, aluminum, zinc)
Silicone resin products, water dilutable (M-SF01)	emulsions of silicone resins mineral fillers, inorganic/organic pigments, water (dispersing agent), additives (film forming agents < 3%)	

Table 10.2 Important coating compounds and applications *(continued)*

qualities are needed in road-marking paints, safety-marking paints, paints used in factories, and paints that render a surface either a good or a poor conductor of electricity.

Metals surfaces may be coated to improve their workability in mechanical processing. Substrates protected from different environmental conditions include the metals (steel, zinc, aluminum, and copper), inorganic materials (plaster, concrete, and asbestos), and organic materials (wood, wallboard, wallpaper, and plastics). Different technical developments have occurred in the coating industry, which permit the use of a variety of raw materials. It is possible to formulate surface coatings that are suitable for each and every kind of material. In many cases a number of different coating systems may come into consideration for painting a particular substrate. In almost all cases a painting problem may be solved in a variety of ways.

Resins are employed in the manufacture of most of the coating compounds. A significant amount of all the resins produced are consumed as coating materials in the various forms. Based on the relatively low capital investments in coating-producing plants, the return on investment is excellent when compared to that of other industries. Net profit margins continue to be in the 6% of sales bracket. At the present, profit for the major type products has been principally due to improved plant manufacturing efficiency. Today's production is by batch process, automation, and mass-production techniques, depending on the quantity of the product and the manufacturing lot's size.

DIFFERENT COATING ASPECT

Coating materials and coating techniques can be distinguished and systematized in various ways (374). The fundamental principles of common coating systems are

1. *Physical drying*. A solid surface film is formed after the evaporation of water or organic solvents.
2. *Physico-chemical drying/curing*. Polycondensation or polyaddition are combined with evaporation of organic solvents.
3. *Chemical curing*. Solvents, such as styrene or acrylic monomers, react with the curing system.

The actual effects depend on the surrounding conditions and the coating system's ingredients, such as solvents. Solvents contribute many essential properties to coating systems. Solvents can improve technical factors such as application or surface properties. They also bring negative qualities to coating materials, especially with respect to environmental conditions (e.g., toxic effects of emitted organic solvents).

Technical application techniques for coatings can be considered in various ways. The stability and durability of coating is essential. Coatings that have normal wear-and-tear requirements are based mainly on oils and aldehyde resins. Higher durability or stability can be achieved by the use of one of the following one- or two-component systems.

Some examples of one-component systems are bituminous material, chlorinated rubber, polyvinyl chloride (PVC), polyacrylic resin, polyethylene, saturated polyester, and polyamide. Some examples of two-component systems are epoxy resin, polyurethane (PUR), and mixtures of reactive resins and tar.

Table 10.3 gives a survey of the performance of different coating materials and an assessment of various environmental factors.

The following are examples of architectural concrete surface coatings. According to DIN 1045 (chapter 22) concrete must be protected against aggressive substances if the chemical attack is severe and long-term. These are the requirements:

Material	Abbreviation	Weathering response	Acid atmosphere	Humidity	Under water	Chemical stress		Solvent	Temperature <60°C	Mechanical	
						acid	alkali			abrasion	scratch
plant oil (linseed oil)	OEL	+y,s,f	+/-	+/-	-	-	-s	-	+	-	-
alkyd resin	AK	+	+/-	+/-	-	-	-s	-	+	-	-
bitumen	B	+/-e,f	+	+	+	+/-	+	-	+/-	-	-
chlorinated rubber	RUC	+/-b,c,f	+	+	+	+	+	-	+	+/-	+/-
PVC-soft	PVC	+/-e,f	+	+/-	+/-	+	+/-	-	+	+/-	+/-
Polyacrylic resin	AY	+	+	+	+	+	+	-	+	+	+/-
polyethene	PE	+/-e	+	+	+	+	+	+	+	+	+
polyester (saturated)	SP	+	+	+/-	+/-	+	+/-(s)	+	+	+	+
epoxy resin	EP	+/-c,g	+	+	+	+/-	+	+	+	+	+
polyurethane	PU	+	+	+/-	+/-	+	+/-	+/-	+	+	+
epoxy resin tar	EP-T	+/-c,e	+	+	+	+/-	+	+/-	+	+	+
polyisobutylene	PLB	+/-	+	+	+	+	+	-	+	+	+

+ = suitable, +/- = limited suitability, - = unsuitable, () = less distinct; b = bleaching, c = chalking, e = embrittlement, f = acid fragments, g = loss of gloss, s = saponification, y = yellowing

Table 10.3 Environmental performance of some coating materials

1. Good adhesion
2. Waterproof and resistant to aggressive substances and resistant to the alkalinity of the concrete
3. Deformable

To realize these requirements, technical solutions such as surface treatments or paint coatings based on thermoplastic (TP) substances can be applied.

Protective coatings based on resins are used in construction with and without fillers and with or without fiber reinforcing materials (chapter 15). The following techniques are generally used:

1. Film-forming paint coatings (brushing, rolling, spraying)
2. Coating (filling, pouring)

During application, the coating materials are normally liquid and subsequently harden by evaporation of solvents or as a result of chemical reactions. The common coating systems for concrete are listed in Table 10.4. Resins such as chlorinated rubbers, styrenes, and acrylics contain normally 40 wt% to 60 wt% solvents and form a thin film. Several coats must be applied. Reactive resins may require little or no solvent.

Wet coatings for metals are listed in Table 10.5.

The drying process does not affect the chemical properties of binders if coatings are of physically drying type. After a solvent has evaporated, the resin molecules become intermeshed, thus producing the desired coating properties.

In contrast to physical drying, binders based on reactive resins, such as epoxy resins, PURs, or polyesters, consist of two components: liquid resin and curing agent. They are either mixed shortly before the coat is applied or, in the case of a one-component system, applied as a slowly reacting mixture. The setting reaction occurs at the surface of the coated material. The final products are normally more resistant and more compact than products based on physically drying binders. Pretreatment of substrate is more critical for applications where chemically hardening products are applied. Coatings can be more or less permeable to water vapor and oxygen. Damage to the metal substrate can occur if water and oxygen reach the reactive surface simultaneously. This is normally impossible if the coatings adhere well and the coated surface is continuous. The adhesion of the coating also prevents penetration by harmful substances via diffusion processes. Adsorption and chemical bonds enhance adhesion.

Other aspects of solvents contained in paint coatings and varnishes are available. The market offers a wide spectrum of coating systems. Examples of common industrial coating materials are listed in Table 10.6 and relevant aspects concerning application and environmental or health risks are also included. Table 10.7 provides examples of release coating systems.

Film-forming agents	Hotmelts	Film-forming agents dissolved	Film-forming agents emulsified	Reaction resin liquid
Bituminous substances	+	+	+	
Svnthetic resin		+	+	
Reactive resins		(+)	(+)	+

Table 10.4 Survey of often-used coating systems for concrete

		Abbreviation	Application	State	Curing method
Natural substances	oil	OEL	binders for wet coating	liquid	oxidizing
	bitumen	B			physical
	tar	T			physical
Plastic materials	alkyd resin	AK	binders for wet coating		oxidizing
	polyurethane	PU			chemical
	epoxy resin	EP			chemical
	acrylic resin	AY			physical
	vinyl resin	PVC			physical
	chlorinated	RUC			physical
	rubber				

Table 10.5 Wet coating materials for metals

TERM AND PERFORMANCE INTRODUCTION

Coatings are generally identified as paints, varnishes, and lacquers. Other nomenclature includes enamels, hot melts, plastisols, organosols, water-emulsion and solution finishes, nonaqueous dispersions, power coatings, masonry water repellents, polishes, magnetic tape coatings and overlays, and so on. There are 100% resin coatings such as vinyl-coated fabrics or PUR floor coverings. The most popular coatings, and the largest user of resins, are paints. Almost all the binders in paints, varnishes, and lacquers are made up principally of resins (Table 10.1).

As reviewed in chapter 1, in the plastics industry, materials reviewed in this book can be identified by different terms such as polymer, plastic, resin, elastomer, and reinforced plastic (RP). They are somewhat synonymous. Polymers, the basic ingredients in materials, can be defined as synthetic or natural high-molecular-weight organic chemical compounds. Practically all of these polymers

Product group/ compounds	Application	Hazardous ingredients, especially solvents	Relevant health and environmental risks
zinc dust coating based on epoxy resin	corrosion protection/primer, application by brush, spray, airless-spraying	xylene 2,5-25%, ethylbenzene 2,5-10%, solvent naphtha 2,5-10%,1-methoxy-2-propanol 1-2,5%	irritations (respiratory, skin, eyes), neurological (narcotic) effects, absorption (skin), flammable, explosive mixtures
reactive PUR-polymers containing solvents	corrosion protection/coating	solvent naphtha 10-20%, diphenylmethane isocyanate 2,5-10%	sensitization (respiratory) irritation (skin, eyes, gastrointestinal tract), neurological effects (narcotic effects, coordination), absorption (skin), flammable, water polluting
modified polyamine containing solvents	corrosion protection/coatings/rigid system	4-tert-butylphenol 10-25%, M-phenylenebis 2,5-10%, trimethylhexane, 1,6-diamine 10-25%, nonylphenol, 10-25% xylene 10-25%, ethylbenzene 2,5-10%	sensitization (skin, respiratory), irritations (skin, eye, respiratory, gastrointestinal tract), neurological (narcotic) effects (coordination), flammable, water polluting
modified epoxy resin containing solvents	corrosion protection/coatings/ rigid system	bisphenol A (epichlorohydrin) 25-50% oxirane, mono(C_{12} -C_{14}-alkyloxy)methyl derivatives 2,5-10%, cyclohexanone 1-2,5%, 2-methylpropane-1-ol 2,5-10%, xylene 10-25%, ethylbenzene 2,5-10%, benzyl alcohol 1-2,5%, 4-methyl-pentane-2-one 1-2,5%	irritations (skin), sensitization (skin, contact), neurological (narcotic) effects, absorption (skin), flammable, water polluting
modified epoxy resin containing solvent	corrosion protection/top layer	bisphenol A (epichlorohydrin) 50-100%, 3-amino-3,5,5-trimethyl ethylbenzene 10-25%, xylene 10-25%	irritation (skin, eyes, respiratory), sensitization (skin), flammable, explosive gas/air mixtures, hazardous reactions (with acids, oxidizers), water polluting
filled and modified epoxy resin	coatings and corrosion prevention/ pore sealer	bisphenol A (epichlorohydrin) 50-100%, P-tert-butylphenyl-1-(2,3-epoxy)propyl-ether 1-2,5%	irritations, sensitization, water polluting
filled and modified epoxy resin	coatings and corrosion prevention/ pore sealer	trimethylhexamethylendiazine-1,6-diamine 10-25%, trimethylhexamethylendiamine -1,6-cyanoethylene 50-100%	irritations, sensitization, water polluting
filled and modified epoxy resin	flooring/mortar screen	bisphenol A (epichlorohydrin) 25-50%, benzyl butyl phthalate 2,5-10%, xylene 2,5-10%	irritation (eye, skin, respiratory), sensitization (contact, skin), explosive gas/ air mixtures (with amines, phenols), water polluting
modified polyamine	flooring/mortar screen	benzyl alcohol 10-25%, nonylphenol 25-50%, 4,4′-methylenebis(cyclohexylamine) 10-25%, 3,6,9-triazaundeca-methylendiamine 10-25%	irritation (eye, skin, respiratory), sensitization (contact, skin), hazardous reactions (with acids, oxidizers), flammable, water polluting

Table 10.6 Examples of coating materials including those containing solvents

filled and modified epoxy resin	corrosion protection/top layer	naphtha 2,5-10%, 2-methoxy-1-methylethyl-acetate 2,5-10%, 2-methyl-propane-1-ol 1-2,5%, xylene 10-25%, ethylbenzene 2,5-10%	irritation (skin, eyes, respiratory), neurological (narcotic) effects, flammable, explosive, hazardous reactions (with oxidizers), water polluting
copolymer dispersion	walls, especially fungicidal properties, good adhesion in damp environment, resistant against condensed water	isothiazolone 1-2,5%, 3-methoxybutylacetate 1-2,5%	irritations (skin, eyes, gastrointestinal tract), water polluting
coating of synthetic resin containing solvents	corrosion protection/coating (steel)	solvent naphtha (petroleum) 25-50 %, xylene 2,5-10%, ethylbenzene 1-2,5%	irritations (eyes, skin, respiratory, gastrointestinal tract), neurological (narcotic) effects, flammable, explosive gas/air mixtures, hazardous reactions (with oxidizers), water polluting
bituminous emulsion (phenol-free, anionic)	corrosion protection/coating (steel, drinking water tanks)	naphtha (petroleum) 25-50%	irritations (skin, eyes, respiratory), neurological effects (co-ordination), flammable, water polluting
modified, filled anthracene oil and polyamine	corrosion protection/top layer	solvent naphtha 2,5-10%, biphenyl-2-ol 2,5-10%, 3-amino-3,5,5-trimethylcyclo-hexamine 1-2,5%, xylene 1-2,5%	irritations (skin, eyes, respiratory), flammable, water polluting
partially neutralized composition of aminoalcohols	sealants and adhesives/elastic products (floor joints, joints)	2-aminoethanol 2,5-10%	sensitization, irritations, slight water pollution
filled, reactive PUR-polymers	sealants and adhesives/elastic products (floor joints, joints)	3-isocyanatemethyl-3,5,5-tri-methylcyclohexylisocyanate 0,1-1%, N,N-dibenzylidenepolyoxy-propylene diamine 1-2,5%, xylene 2,5-10%, ethylbenzene 1-2.5%	sensitization, irritations, hazardous reactions (with amines, alcohols)
solvent based composition, based on PVC	coatings and corrosion prevention/impregnation, sealer	solvent naphtha 50-100%, 1-methoxy-2-propanol 2,5-10%,	irritations, flammable, neurological effects
acrylate dispersion, water dispersed adhesion, promotor	coatings and corrosion protection/ rigid coat (for concrete and dense mineral substrates)	water-borne so-called solvent-free systems	irritations (long-term contact), slightly water polluting

Table 10.6 Examples of coating materials including those containing solvents *(continued)*

Type of System	Cure System	Catalysis	Key Benefits/Features	Applications
Solventless	Addition	Platinum Cure	Cost savings from faster machine speeds; easier coating of temperature sensitive substrates; wide range of release values possible for PSAs; double-sided coatings possible; low temperature cure; no solvent emissions; adjustable release levels; nonblocking.	Release liners for PSA laminates; nonstick food packaging; casting papers; industrial release papers; release films.
Solvent Dispersion	Addition Cure	Platinum	Absence of post cure allows in-process control; cost savings from faster machine speeds; accelerator allows coating of temperature sensitve substrates; stable release values enhance quality control; solvent recoverable; low oven dust.	Release liners for PSA laminates and tapes; casting papers; industrial release papers and films; two-sided differential release papers and films; non-stick food packaging.
Emulsion	Addition Cure	Tin	No fire hazards from solvents; no solvent recovery costs.	Baking papers; nonstick food packaging.
Emulsion	Condensation Cure	Tin	No fire hazards from solvents; no solvent recovery costs; many low release applic-ations; low coating weights possible with thickeners; no solvent emissions; low re-lease values; may be used with thickeners to lower coat weight.	Industrial release papers; casting papers; base papers for PSA laminates; nonstick packaging; water resistant board;
Solvent Dispersion	Condensation Cure	Tin	Wide selection of system components with many combinations possible adjustable release levels; low inhibition; history of documented performance.	Release liners for PSA laminates and tapes; casting papers; industrial release papers; food grade release and packaging papers.
Solventless	Addition Cure	Rhodium	No gelling problems; larger batches poss-ible; broad range of possible release values; no adhesion problems in the roll; consistent product quality; long bath life; adjustable release levels; nonblocking; no solvent emissions; low inhibition.	Release liners for PSA laminates; casting papers; industrial release papers.
Emulsion	Addition Cure	Platinum	No fire hazards from solvents; no recovery costs; permits on paper machine use; long bath life; fast, low-temperature cure; no solvent emissions; good shear stability.	Food grade release and packaging papers; industrial release papers; select pressure sensitive release applications.

Table 10.7 Typical release coating systems and applications

are compounded with other products (additives, fillers, reinforcements, etc.) to provide many different properties and processing capabilities or both; coatings are an important example of this kind of polymer. Thus *plastics* is the correct term to use except in very few applications, where only the polymer is used to fabricate products. However, in the coatings industry, the term *resin*, not *plastics*, is the more commonly used term.

The term *paint* is often used nonspecifically to cover all the coating categories as though the term was synonymous with coating; the terms are often used interchangeably. Paint coatings consume by far the largest quantity of coating materials. However, the other coating processes are important and useful. All these surface coatings represent a large segment of the plastic and chemical industries.

PAINT

Paint consists of three main components, namely, the binder (resin), the pigment, and the solvent. The function of the binder is to provide the forces that hold the film together (cohesive forces) and that hold the film and the substrate together (adhesive forces; 374).

The pigment is a fine powder whose function is to give a coating its desired color and hiding properties. Pigments have a considerable influence on the consistency of the paint and in turn on its application properties. Pigments are also of importance for the resistance of the coating to external attack, in that they are partially responsible for such properties as hardening and resistance to abrasion and weathering.

The solvent is a volatile liquid whose function is to dissolve such binders as would be solid or semisolid at normal temperature. In addition to these three basic components, modern coatings may contain additives of various kinds. Examples are plasticizers, dryers, wetting agents, flattening agents, and emulsifiers or other stabilizers.

The binder is the most important of the three main components and is always present in a manufactured paint. It usually represents 40 wt% to 50 wt% of the paint. Many of the properties of paints and related products are determined directly by the nature of the binder. For this reason paints are often classified, and may even be named, according to the type of binder (Table 10.8).

Some binders are identified or arranged according to the type of drying. A differentiation is made between physical and chemical drying in accordance with the way a coating forms, such as the following:

1. Physical film formation (evaporation of solvent or of dispersion medium in the case of lattices) that includes cellulosics (e.g., nitrocellulose and other esters of cellulose and ethyl cellulose), vinyl resins (e.g., PVC, polyvinyl acetate [PVAc], and polyvinyl acetal), acrylic ester resins, chlorinated rubber, and natural resins (e.g., shellac, rosin, and rosin ester [ester gum]; bitumen [asphalt]; and glue)
2. Chemical film formation (convertible, oxidative drying) that includes drying oils, linseed oil, tung oil, varnishes and oleoresins, and alkyd resins modified with drying oils
3. Cold curing that includes urea-formaldehyde resins, unsaturated polyester resins, epoxy resins, amine-cured resins, and PUR resins
4. TS curing that includes short or medium oil length alkyd resins modified with nondrying oils, water soluble alkyds, epoxy resins cross-linked with amino or phenolic resins, water-soluble addition polymers cross-linked with amino or phenolic resins, and acrylic resins cross-linked with amino or phenolic resins

Curing may be defined as a process in which drying occurs by a chemical reaction between the molecules of the binder without the involvement of gaseous oxygen. If the reaction occurs at room temperature the products are described inaccurately as "cold curing lacquers." If temperatures of 70°C (158°F) or higher are necessary to cause rapid reaction, the materials are known as staving or baking coatings. In view of the many different kinds of chemical reactions that are now used to produce insoluble coatings, the term *convertible coating* is used.

A convertible coating may be defined as one in which the final form of the binder, in the film differs chemically from the binder in the form in which it is applied. The conversion of one form to

Coloring agents
- soluble pigments
  - natural pigments
  - synthetic pigments
- insoluble pigments
  - inorganic pigments
  - organic pigments
    - animal and vegetable pigments
    - synthetic pigments

Binders
- water dilutable binders
  - slaked lime (lime colors)
  - standard cement (cement colors)
  - sodium silicate (colors of 1 or 2 components)
  - vegetable glues (limewash)
  - casein (lime-casein products, alkali-lime products)
  - dispersions
    - natural resin emulsion paints
    - plastomer emulsion paints (PVAC (homopolymers, copolymers), PVP, polyacrylates (PMMA, styrol-acetate))
    - water emulsifiable varnish systems (aqueous acrylate systems, aqueous polyurethane systems)
- solvent dilutable binders
  - oil paints
  - varnishes
    - products drying by air oxidation (nitrocellulose varnishes, aldehyde resin varnishes, oil varnishes)
    - physical drying products (polymer resin varnishes, polyvinyl chloride varnishes, polyvinyl acetate varnishes, polyacrylate varnishes, chlorinated rubber varnishes)
    - chemical curing products (phenolic varnishes, aminoplast varnishes, melamine resin varnishes)
    - acid-curing varnishes
    - epoxy resin varnishes
    - polyurethane varnishes
    - unsaturated polyester varnishes

Table 10.8 Example of paint and varnish coating compositions

the other may be achieved by the action of some component of the atmosphere, such as oxygen or water; by heat; by radiation; by the use of catalysts; or by a reaction between two or more binder components that are mixed just prior to application.

These reaction-type coatings provide films with greater hardness and chemical resistance than those obtained by oxidative drying.

WATER-BASED PAINT

Water-based, *water-thinned*, *aqueous*, and other terms are used to refer to paints that contain water. Technically, three types exist: (1) latex or emulsion paints made with synthetic resins such acrylic, PVAc, or butadiene-styrene; (2) water-soluble oils or alkyds; and (3) emulsified oils or alkyds.

Water-based paints using casein and emulsion oil paints containing alkyd resin and water were first introduced just prior to World War II. Latex paints using butadiene-styrene followed after World War II. These were rubber-based paints that lacked ruggedness. In 1953 the acrylic emulsion type of paint was introduced for indoor surfaces and outside masonry surfaces. By 1957, acrylic emulsion types for exterior wood surfaces were on the market.

Water-based coatings continue to gleam in industry's eye. Elimination of solvent fumes from these systems reduces fire and explosion hazards, improves working conditions, and lowers insurance rates. These systems are more expensive in terms of both the coating and the "paint booth apparatus." Water is more costly to evaporate, and its rate of evaporation is more difficult to control.

VARNISH

The word *varnish* first appeared during the sixteenth century. It denoted a fluid mixture of amber and oil, or more generally, of resin and oil. This latter meaning has survived to the present day.

LACQUER

The term *lacquer* is frequently applied to almost any coating composition that dries solely, and rapidly, by evaporation of the solvent. It originally was almost exclusively associated with nitrocellulose-based coatings. At the present time it generally refers to coatings that contain nitrocellulose or possibly another cellulose derivative.

SOLVENT

The coating and other industrial processes include relying on the dissolution of raw materials and subsequent removal of solvents by various drying processes. The formation of a solution and the subsequent solvent removal depend on solvent transport phenomena that are determined by the properties of the solute and the properties of the solvent. A solvent is a material, usually a liquid that has the power to dissolve another material and form a homogeneous mixture known as a solution (Table 10.9). Most of these are toxic and flammable so exercise caution when using them.

Knowledge of a solvent's movement within the solid matrix by a diffusion process is essential to design the technological processes. Many of the final properties, such as tribological properties; mechanical toughness; optical clarity; protection against corrosion adhesion to substrates and reinforcing fillers; protective properties of clothing; quality of the coated surface; toxic residues; morphology and residual stress; ingress of toxic substances; and chemical resistance depend not only on the material chosen but also on the regimes of technological processes. For these reasons, solvent transport phenomena are of interest to the modem industry (374).

Thin film coating and drying technology are the key technologies for manufacturing diverse kinds of functional films, such as photographic films, adhesives, image media, magnetic media, and lithium battery coating. Coating applied to a substrate as a liquid needs some degree of solidification

Solvent	Characteristics
Lacquer Thinner	A mixture of toluene, isopropanol, methyl isobutyl keytone, acetone, propylene glycol, monomethyl ether acetate and ethyl acetate. Photochemically reactive. Used to thin lacquers and epoxies but can be used as a general cleaner and degreaser also highly flammable. Dissolves or softens many plastics.
Acetone	2–Propanone or Dimethyl ketone is the actual chemical, CH_3COCH_3, soluble in water and alcohol, non-photochemically reactive: used to clean and remove epoxy resins, polyester, ink, adhesives, contact cement, and fiberglass cleanup. Dissolves or softens many plastics.
Finger Nail Polish Remover	Mixture of acetone, cocamidopropyl dimethylamine propionate, and amp isostearic hydrolyzed animal protein. Good for various apps. Dissolves plastic.
Plastic Cement	Methyl ethyl keytone or 2–Bufanone, $CH_3CH_2COCH_3$, soluble in water and alcohols, actually a solvent that dissolves plastic and is typically used in making model airplanes.
Methylene Chloride	Dichloromethane, CH_2Cl_2, not a very common solvent but when mixed with xylene (dimethylbenzene) makes a strong solvent for things like crayon marks, lipstick, ink, magic marker, gum, latex, oil, and wax. Dissolves or softens some plastics. Marketed as "Klean-Clean" by Klean–Strip. W.W. Barr Inc, Memphis, TN.
Naphtha	Slight smell but good for some applications. Non–photochemically reactive. Very fast evaporation.
Turpentine (Paint thinner)	"Steam Distilled" or "Gum Spirits" made from pine trees: used as a thinner and cleaner for oil based paint, varnish, enamel and stain. Photochemically reactive.
Solvent Alcohol	Methanol or Methyl Alcohol or wood alcohol. CH_3OH non-photochemically reactive, poisonous, used primarily as a thinner for shellac base primers. Do not use with oil or latex paints, stains, or varnishes. Also used in manne alcohol stoves, soluble in water and other alcohols. Can be mixed with gasoline in the gas tank to eliminate moisture problems (1.2 pint per 15 gallons). Good cleaner for computer plastic parts.

Table 10.9 Examples of solvents and their behaviors

Freon®	Trichlortnifluoroethane, Freon® TF, available in spray cans, non-flammable, non-conductive, low toxicity, odorless and does not attack plastic, rubber, paints or metal; low surface tension, evaporates fast. Although not commonly seen, freon is an excellent solvent and is typically used to clean electrocal connectors and computer components. Note: Depletes ozone, must be certified.
Denatured Alcohol	Ethyl or grain alcohol made unfit for drinking by the addition of compounds Soluble in water, and other alcohols. Non-photochemically reactive, typically used to thin shellac, clean glass and metal, to clean ink from rubber rollers, and as a fuel in marine stoves. To clean glass porcelain and piano keys, mix 1.1 with water.
Rubbing Alcohol (isopropyl)	2–Propanol is the actual chemical CH_3CHOCH_3 soluble in water and other alcohol, general cleaner and disinfectant specifically used to clean tape recorder heads and computer disk drive heads.

Table 10.9 Examples of solvents and their behaviors *(continued)*

in order to be part of a final product. The degree of solidification can be low in the case of pressure-sensitive adhesives (PSA), and it ranges to high in the case of dense metal oxides. The final structure and properties of coating are greatly influenced by the drying conditions. Poorly chosen operating conditions of drying cause unwanted internal gradients, phase separations, colloidal transformations that lead to the wrong microstructure, inappropriate nonuniformities, and stress-related defects.

Solvent removal, or drying, which is part of the solidification process, is an important subject. Typically, a coating solution consists of pigments, binders (plastics), and solvents. The solvents are used to make the coating formulation soluble and to give the coating solution (or dispersion) the rheology necessary for the application. The coating solution is deposited onto a substrate or web at the coating station and is dried by passing through a series of separate ovens (zones). A substrate can be an impermeable material such as plastic film or a permeable material in the case of paper coating. The dryer consists of ovens (zones) in which the temperature and velocity of air are controlled independently.

Figure 10.1 is an example of an industrial coating and drying apparatus. A coated liquid is deposited onto a substrate, which is unwound from a supply roll at the coating station and passes through three separate ovens (dryer) in which the temperature and velocity of air are controlled independently. Finally a take-up roll takes up the dried, coated substrate. The basics of the process of drying are shown in Figure 10.2.

The air impinges on the coated and backside surface of the substrate through the nozzles and sweeps away the solvent vapor from the coated surface. In the case of a single-sided impingement

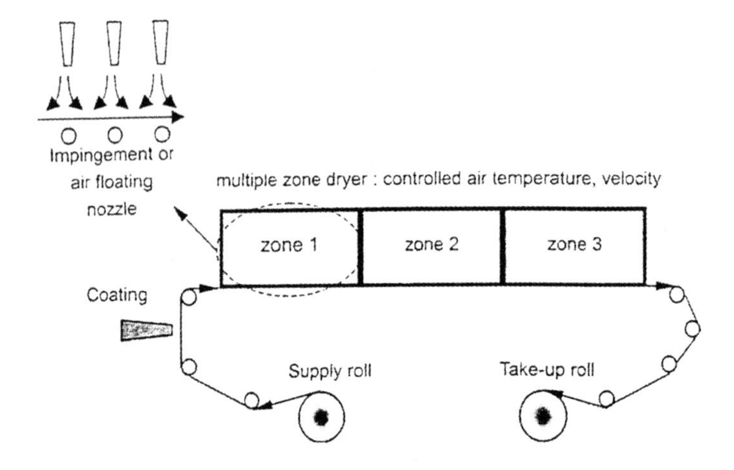

Figure 10.1 Example of industrial coating and drying apparatus.

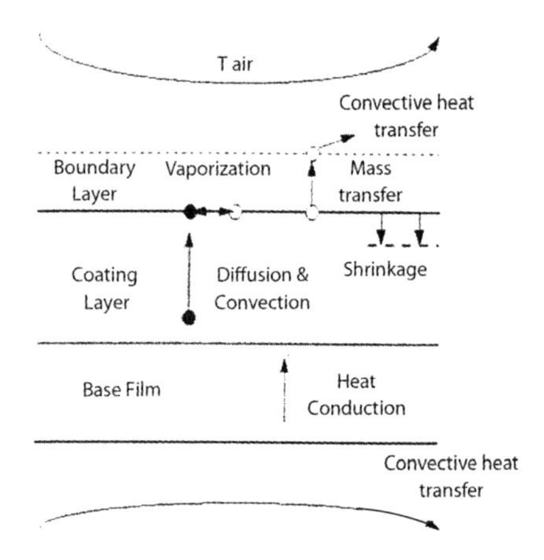

Figure 10.2 The basic drying process and typical drying parameters.

dryer, the air impinges only on the coated surface. The coated film must be dry before reaching the rewind station. The residence time of a coated substrate in the dryer is as short as several seconds in the case of high-speed magnetic coating processes, and it ranges to as long as several minutes in a case of lithium battery–coating processes. Finally, a take-up roll takes up the dried, coated substrate.

Free-radical polymerization is one of the most useful and lucrative fields of chemistry. In recent years there has been a tremendous increase in research in this area, which was once considered a mature technological field. Free-radical synthetic polymer chemistry is tolerant of diverse functionality and can be performed in a wide range of media.

Emulsion and suspension polymerizations have been important industrial processes for many years. More recently, the "green" synthesis of resins has diversified from aqueous media to supercritical fluids and the fluorous biphase. An enduring feature of the research literature on free-radical polymerization has been studies into specific solvent effects. In many cases the influence of solvents is small. However, it is becoming increasingly evident that solvent effects can be used to assist in controlling the polymerization reaction, at both the macroscopic and the molecular levels (374).

PROPERTIES OF PLASTICS

Almost all the binders in paint, varnish, and lacquer coatings are composed of plastic materials. The plastics are applied in one operation or built up during the drying processes. For example, the physical and chemical properties of vinyl coatings have a direct relationship to the basic polymeric material. The use of synthetic resins dates back to the turn of the twentieth century. Up until World War I they were principally used as low-cost substitutes for natural resins. Since 1915 many different plastics that offer many advantages compared with the natural resins have been used (Table 10.10).

The synthetic resins are less subject to variations in availability and consequently have more stable costs. They may be produced to fairly close technical tolerances, while the natural resins show wide variations in quality. More important, however, is the fact that the synthetic resins can be varied in relation to the end use for which they are intended.

The major long-range trend in paints and related surface coatings is toward greater efficiency. The target continues to be products with better environmental protection or decoration for longer periods of time at lower total cost per square foot. Paints compete with a variety of other surfacing materials, such as wallpaper (vinyl, polyethylene, and polyvinyl fluoride [PVF] films); porcelain

Type of coating	Flexibility	Chemical resistance	Stain resistance	Exterior durability
Vinyls (solution)	Excellent	Excellent except for solvents	Limited	Very good
Alkyds	Fair	Limited	Good	Good
Plastisols and organosols	Excellent	Excellent except for solvents	Good	Very good
Nitrocellulose	Poor	Fair	Very good	Good
Epoxies	Excellent	Excellent	Very good	Poor
Phenolics	Poor	Excellent	Excellent	Poor
Vinyl-alkyds	Very good	Good	Limited	Good
Acrylics	Limited	Good	Excellent	Very good
Fluorocarbon	Excellent	Excellent	Excellent	Excellent
Silicones	Excellent	Very good	Very good	Excellent

Table 10.10 Examples of coating performances

enamels; and electroplated, phosphated, or oxidized metal films. In addition, coatings and their substrates compete against structural materials that require no special surface coatings, such as stainless steel, aluminum, glass, stone and brick, RPs, extruded plastics, and molded plastics. However, there are many new applications in which materials such as steel, aluminum, wood, concrete, and brick are given plastic coatings to provide more durable and attractive products.

To meet competition, the paint industry continues to develop new formulations and new methods of applications. Since the best performance per unit cost is desired, there is continual effort to lower the cost per square foot per surface coated, either by lowering material costs per pound, using thinner films, or devising more economical means of application. Application techniques have involved extensive laboratory tests with different composite resins.

THERMOPLASTIC COATING

Serviceable TP films must have a minimum level of strength that depends on the end use of the product. Film strength depends on many variables, but a critical factor is molecular weight (MW). MW varies with the chemical composition of the plastic and the mechanical properties required for the application. For example, in solvent evaporation from solutions of TPs for spray applications, consideration has to be given to the solvent's evaporation behavior. With a methyl ethyl ketone (MEK) solvent and a vinyl copolymer that has a relatively high vapor pressure under application conditions, MEK evaporates quickly.

With this type of system, a large fraction of the solvent evaporates in the time interval between the coating leaving, as an example, the orifice of the spray gun and its deposition on the surface being coated. As the solvent evaporates, the viscosity increases and the coating reaches the dry-to-touch state soon after application and does not block. However, if the film is formed at 77°F (25°C), the dry film contains several percent of retained solvent.

In the first stages of solvent evaporation from such a film, the rate of evaporation depends on the vapor pressure at the temperatures encountered during evaporation, the surface area-to-volume ratio of the film, and the rate of airflow over the surface. It is essentially independent of the presence of plastic. The rate of solvent diffusion through the film depends not only on the temperature and the glass transition temperature (T_g) of the film, but also on the solvent structure and any solvent-plastic interactions. The coating thickness is another parameter that affects solvent loss and film formation.

TP-based coatings have a low solids content because their relatively high MWs require large amounts of solvent to reduce the viscosity to levels low enough for application. Air pollution regulations limiting the emission of volatile organic compounds (VOCs) and the increasing costs of solvents have led to the increasing replacement of these coatings with lower-solvent or solventless coatings. However, large-scale production means that solvent-coating systems become economically beneficial when used with a solvent-recovery system.

LATEX

Emulsions of latex (or paint) have low solvent emissions as well as other advantages. Latex is a dispersion of high-MW plastic in water. Charge repulsion and entropic repulsion (also called steric repulsion and osmotic repulsion) stabilize the dispersion. Because the latex plastic is not in solution, the rate of water loss is almost independent of composition until the evaporation gets close to its end. When a dry film is prepared from latex, the forces that stabilize the dispersion of latex particles must be overcome, and the particles must coalesce into a continuous film. The rate of coalescence is controlled by the free volume available, which in turn depends mainly on T_g. The viscosity of the coalesced film also depends on the free volume.

With latex or paint emulsion, coating material is made of two dispersions: (1) dry powders with colorants, fillers, and extenders and (2) plastic dispersions. These emulsion paints have the binder in a water-dispersed form. Principal types are styrene-butadiene, PVAc, and acrylic plastics. Percentage composition by volume is usually 25% to 30% dry ingredients, 40% latex, and 20% water plus stabilizer. Their unique properties are ease of application, absence of disagreeable odor, and nonflammability; they are used both indoors and out.

TS COATING

A potential problem in TS plastic systems is the relationship between storage stability of the coating before application and the time and temperature required to cure the film after application. The processing of TSs is different than that of TPs (chapter 1). It is desirable to store a coating for many months or even years without a significant increase in viscosity caused by cross-linking reaction during storage. However, after application, the cross-linking (cure) should proceed in a short time at as low a temperature as possible.

Since reaction rates depend on the concentration of the polymer's functional groups, using more dilute systems can increase the storage life, which is achieved by adding more solvent. When the solvent evaporates after application, the reaction rate will initially increase. Although it is advantageous to reduce solvent concentration as much as possible, the problem of storage stability has to be considered for systems with a higher solids content.

The mechanical properties of the final film depend on T_g for the cross-linked polymer and the degree of cross-linking, or the cross-link density (chapter 1). The average functionality, the equivalent weight of the system, and the completeness of the reaction (complete cure of the TS) affect the cross-link density.

FUNDAMENTALS OF RESIN FORMATION

There are two classes of resins: they can be identified as condensation or addition types (chapter 1). Condensation polymerization is the process by which a polymer is built up by successive reactions between monomer molecules and the growing polymer. In each reaction step, condensation polymerization produces a small molecule such as water, hydrogen chloride, or sodium chloride and increases the polymer size.

Addition polymerization is the process by which a polymer is built up by a repeated addition reaction between monomer molecules and the growing polymer. This action occurs within any other reaction product when the polymer is being formed. The monomer, in the majority of cases of practical importance, is an unsaturated compound, usually a vinyl derivative. While an addition polymer has the same elementary composition as the monomer, this is not true for condensation polymers. A review of some of the resins used in coating follows (chapter 2).

CONDENSATION TYPE

ALKYD RESIN

Many variations in the constituents and portions of the alkyd coating material are available. Many different binders, such as drying oils, phenolic resins, amino resins, nitrocellulose, maleic resins, chlorinated rubber and cyclized rubber can be used. As a group, the alkyds are distinguished by rapid drying, good adhesion, elasticity, resistance to marring, and durability.

Their principal weakness resides in the facility with which the ester groups, which form a large part of the molecules, are hydrolyzed (particularly under alkaline conditions). Even in this respect it is possible to produce alkyds with greatly improved resistance to hydrolysis through the use of polyols.

Alkyd resins are instrumental in the coatings used in producing automobiles, refrigerators, washing machines, and many other consumer goods. Styrenated alkyds (in contrast to the styrenated oils) have been used with a fair amount of success as binders in very rapid air drying and rapid, low-temperature stoving finishes. Alkyds are also modified with vinyl derivatives such as esters of acrylic and methacrylic acids, or with mixtures of these compounds and styrene or vinyl-toluene. It appears that none of these combinations has experienced practical commercial success.

POLYESTER RESIN, UNSATURATED

The unsaturated polyesters are of particular interest in the coatings field. The monomer may be used to adjust the viscosity of the coating to the required value. In most cases this action occurs in conjunction with small amounts of solvents. It has had limited successful use in the coating industry principally due to the fact that the curing is strongly inhibited by atmospheric oxygen. The result is that the surface of the polyester coating remains soft and sticky.

PHENOLIC RESIN

The first phenolic resin appeared on the market in 1902. It was a spirit soluble, nonhardening Novolac type. It was intended as a substitute for shellac and spirit varnishes. In 1907, Baekeland's historic patent for the preparation of phenolic resin molding compound was published. This type of phenolic resin was not suitable for coatings. The first patentfor oil-soluble phenolic resin was issued in 1913. There followed different patents for different phenolic resin coating formulations. Many different types are now available with extensive service life.

AMINO RESIN

A number of resins containing nitrogen are classed together as amino resins. This terminology tends to be confusing but continues to be used. Amino resins are obtained by condensation of amino or amido compounds with aldehydes. The most important are the urea and melamine resins (thermosetting) and the aniline resins (TP; chapter 2). The thermosetting coatings are of interest in the coatings field.

UREA RESIN

Urea resins are not used alone as binders and coatings. When they cure, the films are brittle and lack adhesion. The usual modifiers are alkyd resins, as well as combinations with nitrocellulose. In the latter case, however, the improved gloss tends to increase yellowing.

MELAMINE RESIN

Melamine is a white, crystalline powder with a high melting point. It differs from urea in that it has very low solubility in water. Melamine resins are prepared in the same way as urea resins, by condensation with formaldehyde. The melamine resins have replaced the urea resins in many applications. The most important use is in combination with alkyd resins. This combination improves resistance to water, alkali, and chemicals. Virtually nonyellowing finishes may be obtained with suitable choices of alkyds.

EPOXY RESIN

Epoxies have provided the surface coatings industry with a wide variety of formulation possibilities. They are used alone or in combination with other plastics. Alhough they are more expensive than other types of binders, their outstanding properties and versatility continue to expand their applications.

Epoxy resins provide good chemical resistance and, in particular, excellent resistance to alkalis, including caustic alkalis. A major asset is their excellent adhesion to many different substrates. Other important properties include exceptional hardness and flexibility.

One of the disadvantages of epoxy resins is that they are not soluble in the lower-cost solvents. Compatibility with other film formers is limited. Finishes based on epoxy resins have a marked tendency to chalk when used outdoors, and their water resistance is not always the best.

They are used in combination with phenolic, urea, and melamine resins, which act as crosslinking agents. Cold-curing coatings with polyamines or polyamides as curing agents are very popular. Air-dried coatings are also popular. They are produced after esterification with unsaturated fatty acids. The product is known as epoxide ester or epoxy ester.

The combination of epoxy resins with phenolic resins can give maximum resistance to chemicals and solvents, as well as adhesion, flexibility, hardness, and abrasion resistance. The most suitable phenolic resins for cold blending are of the butylated resole type.

Polyamide (nylon) resins, which are formulated to contain free amine groups, can serve as catalysts for epoxy resins. The polyamide resins are practically nontoxic and nonirritant to humans, whereas some of the amine catalysts (used with epoxies) must be handled with special precautions. The polyamide combinations produce tough films in combination with a lower resistance to solvents and chemicals.

Polyurethane resin

There are many different types of urethane (e.g., PUR) coatings on the market. For coating applications, a cross-linked film is preferred and thermoplastic urethanes are of little interest. The usual hazards associated with isocyanates are applicable in preparing these coatings.

With a suitable choice of components, it is possible to obtain almost any degree of flexibility and hardness, ranging from highly elastic films for coating rubber and leather articles to extremely hard, abrasion-resistant coatings for floors, boats, and metal objects. These coatings are important in the coating industry.

Silicone resin

Silicone resins are heat-convertible and are used either alone or in combination with other binders in coatings. Their most important and distinguishing property is resistance to degradation when exposed to high temperatures. In addition, they have good electrical properties and outdoor durability. Suitably pigmented silicone coatings will withstand temperatures of 260°F (127°C) continuously, while most other coatings would not even survive long exposure at 150°F (66°C).

Silicone resins pigmented with aluminum powder or zinc dust give films with good weather and corrosion resistance at temperatures as high as 500°F (260°C). Silicone finishes in the electrical industry provide a combination of heat resistance and electrical insulation.

Addition Type

Polyethylene resin

The important and significant properties available with polyethylene binders are flexibility and water and chemical resistance. However, since they are insoluble in all organic solvents at temperatures below about 50°F (10°C), they are not used in normal surface coatings. They can be applied directly from the solid by flame spraying. In most applications the substrate has to be pretreated in order to provide suitable adhesion.

There are polyethylene compounds that are soluble in either urea or epoxy resins to provide different types of coatings. These are constituted with chloro- and chlorosulfonyl groups. These coatings make highly elastic films characterized by particularly good resistance to strong acids such as concentrated hydrochloric and sulfuric acids, and to oxidizing agents such as ozone, hydrogen peroxide, and chromic acid. They are not resistant to hot concentrated nitric acid.

An example of these special polyethylene compounds is DuPont's Hypalon, which is combined with other binders such as chlorinated rubber, urea resins, and epoxy resins to provide different types of coatings. These different coatings are used in various chemical bath containers.

There are many different polyethylene coating compounds to meet many different coating requirements. ExxonMobil Chemical has a new family of linear low-density polyethylene (LLDPE) extrusion coating resins. The first of a new family of very low-density linear polyethylenes are produced with the company's Exxpol metallocene technology for extrusion coating and laminating. These new Exxco products are said to have "plastomer-like" properties and to significantly outperform conventional extrusion-coating resins in sealing ease, hot-tack and seal strength, tear strength, puncture resistance, and adhesion to stretched oriented film.

Recommended uses involve blending with about 20% LDPE or ethylene-vinyl acetate (EVA) for coating paper and paperboard. The initial grade, Exxco 012, has 0.912 g/cc density and 12 melt flow index (MI). It is a close cousin to ExxonMobil's new Exceed 1012CA metallocene VLDPE film resin of the same density and MI. However, Exxco is aimed specifically at coating and laminating. ExxonMobil is exploring future Exxco grades with higher and lower MIs.

VINYL RESIN

The principal vinyl resins used in coatings are copolymers of vinyl chloride and vinyl acetate. Polyvinylidene chloride (PVDC) and polyvinyl butyral (PVB) are also important. PVAc in emulsion form is widely used in architectural coatings.

The vinyl copolymers produce air-drying coatings that have excellent toughness and good resistance to water and chemicals. However, they are sensitive to heat, ultraviolet radiation, and many solvents. They are high-MW polymers and therefore require fairly strong solvents. Development of the dispersion type of vinyl resin permits their application as organosols and plastisols at high solid content, which extends their usefulness considerably. They do not have high solids at spraying consistency.

Vinyl resins are widely used as fabric coatings because of their combination of toughness and flexibility, and their property of not supporting combustion. Because they are nonflammable, they have replaced nitrocellulose lacquers for many applications on fabrics.

Vinyls produce excellent coatings on metals, but care must be taken in their application because, like most high polymers, they have strong cohesive forces that may overcome the adhesive forces. The entire coating may flake off as a continuous sheet if the precise application conditions have not been complied with for the various modifications.

The absence of odor, taste, and toxicity in vinyl coatings makes them suitable for the lining of beer cans. They have other applications in food containers but certain limitations exist, namely, poor adhesion and sensitivity to temperatures reached in processing foods.

The vinyl copolymers can be used most efficiently in special applications such as hospital and dental equipment, a field in which durability is more important than initial cost. For laboratory equipment, epoxy resins may be preferred because the vinyls are sensitive to some solvents. Vinyl coating systems consisting of corrosion-inhibiting primer and chemical-resistant finish coats are used on new equipment for chemical plants. Metal conditioners based on zinc chromate and PVB are widely used instead of sand-blasted steel on both industrial and marine equipment.

PVAc in the pure and solid form is colorless and transparent. It is somewhat brittle unless the degree of polymerization is low. Its softening temperature is between 40°F and 90°F (4°C and 32°C), depending on the MW. It exhibits the phenomenon of cold flow.

Because of its water solubility, polyvinyl alcohol (PVAL) plays a relatively small part as a binder in surface coatings. It has been used as an impregnant in the production of grease-proof paper, as a yarn sizing, and for the production of water-soluble packages. It is useful as a dispersing agent and a protective colloid, for example, in latex paints. It has an advantage over glue and casein in that it is much less susceptible to microbiological attack.

Dispersion coating, PVC

At this time, most dispersion coatings available on the market are based on the PVC homopolymer type of resin. Other types of dispersion coatings, such as those based on PVF and polyvinylidene fluoride (PVDF), are also available in the marketplace. Major reference is made to the PVC types of dispersion coatings. They have important applications in industrial finishes because of their economy and excellent performance characteristics.

The dispersion technique provides the advantage of the good properties of high-molecular-weight vinyl chloride resins. Dispersion coatings are also known as organosols and plastisols. There are also conventional solution vinyl coatings that perform well for the organic coatings industry. These solution vinyl resin coatings are based on copolymers of vinyl chloride and vinyl acetate and are of relatively low molecular weight. Polymers containing a third component are also used and provide the industry with vinyl polymers that have carboxyl (–COOH) or hydroxyl (–OH) groups, or are otherwise terminated for the attainment of special properties (Table 10.11).

Organosol and plastisol

It has long been known that the higher-molecular-weight vinyl resins produced films that gave the best toughness and resistance properties. The resulting coatings, however, had poor adhesion to metal substrates, gave very low solids when dissolved in even the strongest solvents, and exhibited poor flow properties. By not dissolving this resin, but dispersing it suitably, this family of coating materials became known as the dispersion coatings.

		Vinyl chloride dispersion	Vinyl solution	Modified acrylic
Film corrosion (ASTM B-287-61)		None	None	None
Film adhesion (Crosshatch)		Excellent	Fair-good	Good
Flexibility				
Conical mandrel		Excellent	Excellent	Good
28 mm screw cap		Excellent	Excellent	Fair
Gardner impact (in. lb)				
Unexposed		Very good	Fair	Good
Water soaked	(High gloss)	Very good	Fair	Good
C	(Low gloss)	Good		
Cass test	(High gloss)	Very good	Fair	Fair
	(Low gloss)	Fair		
Tabor abrasion	(High gloss)	.0157	.0112	.0280
(g loss/200 cycles-CS-10 Wheel—1000 g wt)	(Low gloss)	.0061		
Pencil hardness	(High gloss)	2H		
	(Low gloss)	4H	3H	4H
Cost/ft²-1.0 mil film (range)		0.75–0.9¢	0.9–1.2¢	0.65–0.85¢

1000 hours' exposure	Vinyl chloride dispersion	Vinyl solution	Modified acrylic
Atlas weather-ometer	Good	Fair-good	Fair-good
Relative humidity resistance			
(100% R.H. at 100°F)	Very good	Good	Very good
5% Salt spray (AST D4117-S7T)	Very good	Good	Fair-good
Water soak (77°F)	Good	Fair-good	Good
12 months south Florida			
45° south			
Resistance to dirt retention	Good	Good	Good–very good
Resistance to chalking	Excellent	Very good	Very good

Table 10.11 General performance comparisons

Vinyl dispersion is a suspension of colloidal-size particles in an organic medium, which is not capable of dissolving the resin at room temperature but exerts some solvating or peptizing effect on the polymer particles. When the organic medium contains volatile solvent, the mixture is called an organosol. Mixtures that do not contain appreciable amounts of volatile thinners are referred to as plastisols. Organosol dispersions normally contain 40% to 80% solids, whereas plastisols contain 90% to 100% solids.

The resin particle diameter range varies from 0.1 to 1.0 μm. Relative viscosity is in the range of 2.50 to 3.0 for the higher-molecular-weight resins and 2.05 to 2.4 for the lower-molecular-weight materials. Relative viscosity, also referred to as viscosity ratio, is defined as the ratio of the viscosity of a 1% resin solution in cyclohexanone at 25°C (77°F) to the viscosity of pure cyclohexanone at 25°C (77°F).

COMPOSITION

In general, dispersion coatings are composed of the ingredients shown in Table 10.12. The liquid phase of the dispersion system—the solvents, the diluents, and the plasticizers—performs multiple functions, such as serving as a wetting aid, dispersing medium, viscosity depressant (thinner), stabilizer, and fusion aid. Because of these multiple roles, these liquid components must be meticulously selected to achieve a dispersion system with proper application and coalescing properties (Table 10.13). Not only must they be perfectly balanced for good flow properties to prevent flocculation or oversolvation and even gelation of the dispersion resin, consideration must be also given to the requirements of and interactions with the other formula ingredients, such as pigments, resins, and other modifying materials.

Dispersion resin(s)	
Pigments	
Plasticizers	
Diluents and solvents	
Stabilizers and crosslinking agents	
Modifying resins	
Acrylic	Oleoresinous
Alkyd	Phenolic
Amino	Polyester
Epoxy	Silicone
Hydrocarbon (incl. polybutadiene)	Vinyl copolymers
Ketone	Other

Table 10.12 General composition of dispersion coatings

Toughness	Excellent
Impact resistance	Very high
Tensile strength	Very high
Chemical resistance	Unaffected by dilute and concentrated acids and alkalis Resistant to organic solvents, alcohols, greases, and aliphatic hydrocarbons. Inert to water and aqueous salt solutions
Abrasion resistance	Excellent
Water absorption	Very low
Moisture vapor transmission	Very low rate
Flammability	Nonflammable
Toxicity	Odorless, tasteless, and nontoxic
Heat sealing properties	Good
Weatherability	Good–excellent
Stability	Good–excellent
Adhesion (unmodified)	Poor
Adhesion (modified)	Poor–Excellent

Table 10.13 Example of advantages using dispersion coatings

One of the defects noticeable in an organosol film with inadequate solvent balance is called mud-cracking. The organosols are prone to this condition because they inherently tend to release solvents very rapidly. In a deposited film, a too-rapid loss of solvent results in volume shrinkage and hence causes mud-cracking. This condition can be avoided by combining a fast cure rate (to quickly coalesce or fuse the tiny resin particles) with a slow solvent system to keep the film mobile for as long as is necessary to obtain optimum film properties.

In some specially developed organosol coating systems, it is practically impossible to design solvent systems that would produce good flow, aid in proper fuse-out of the film, and still be viscosity-stable on storage. Such coating materials are sold as two-package systems. The organosol component contains a balanced solvent system for the ingredients contained therein, yielding a storage-stable liquid that may be clear or pigmented. The catalyst component, which may contain the modifying resins or cross-linking agents, also utilizes solvents that are properly balanced for this component and it, too, forms a storage-stable liquid, clear or pigmented, as the case may be.

The organosol and catalyst components are usually mixed equal parts by volume to form a coating with the desired end properties. In approximately six hours (depending on specific composition, ambient temperature and humidity), oversolvation may start and viscosity may begin to rise. In such a case, a fresh, equal-parts mixture is added, and the blend may normally be used with complete satisfaction. In commercial practice, the makeup or the replenishing coating material effectively eliminates any practical concern regarding a rise in viscosity. Any catalyzed mixture held over a weekend, for example, is merely checked for fluidity and added to a fresh, equal-parts mixture.

When storing dispersion systems, one should recognize that the dispersants or polar solvents are more powerful in their action at higher temperatures. Thus to avoid oversolvation, organosol materials should not be stored in the hot sun, next to radiators, or in places where the coating temperature may rise to higher than 120°F (49°C). Since solvent power falls off with the reduction of ambient temperatures, undersolvation and possibly flocculation may occur if organosols are stored outside for long periods during the winter in cold areas. Never heat up dispersion coatings rapidly with devices such as immersion heaters or by oversolvation because gelation can occur.

METHOD OF APPLICATION

Dispersion resins can be applied by a variety of methods including spraying, knifing, rolling with a roller, dipping, and extrusion. While spraying techniques are mostly reserved for organosols and extrusion procedures for plastisols, the other methods are common to both types of system. The choice of method of application or dispersion system (organosol or plastisol) is dictated by film thickness requirements, available application techniques (spray, extrusion) and processing equipment (shear mixers or roller mills), substrates, and product performance. It is interesting to note that, due to the puffy or thixotropic nature of dispersion coatings, these materials can be applied at much higher than normal viscosities. The shear forces exerted during recirculation and by the nap of the roller coater, or during spraying or extrusion, are effective in reducing the actual coating viscosity of these materials. In addition, the normal coalescence or fusion of the resin particles in the baking oven assists flow.

Thixotropy is a property of a plastic that is a gel at rest but liquefies upon agitation and loses viscosity under stress. Liquids containing suspended solids are likely to be thixotropic. They have both high static shear strength and low dynamic shear strength. For example, these materials have the capability to be applied on a vertical wall and through quick curing action remain in position during curing.

All dispersion coatings must be properly baked or fused in order to coalesce the tiny dispersed resin particles into a continuous, tough, and flexible film. Depending on formulation and dwell time, the required fusing temperatures (based on actual metal temperatures) vary from 300°F to 525°F (149°C to 275°C). The preferred cycle for sheet bakes is 10 minutes in the 350°F to 525°F (177°C to 275°C) range. In moving-web application (coil or strip coating), a cure cycle of 60 seconds or less at about 525°F yields good results. These cited examples, of course, are for vinyl chloride dispersion systems. Fluorinated dispersion coatings require substantially higher temperatures (approximately 550°F to 600°F [288°C to 316°C]) for proper film formation.

Since fusion of the dispersed particles is the major objective in the curing procedure, the baking cycle for a given application depends on how quickly the wet film reaches fusing temperature. When this critical temperature is reached, the tiny, partially solvated particles quickly coalesce into a homogeneous coating. Problems of thermal degradation will occur if the coating is subjected to temperatures in excess of 500°F (260°C) for vinyl chloride dispersion; degradation will occur at temperatures in excess of 600°F (316°C) for vinyl or vinylidene fluoride dispersions for prolonged periods.

In the instances of vinyl chloride dispersion, the rate of thermal decomposition is accelerated in the presence of iron; such situations occur, for example, when microscopically exposed tin plate is subjected to extreme temperatures for only a few seconds. The resulting film is discolored black. Discoloration of this type can also mean that the oven has been set too high or has developed a hot spot or another similar problem. To reduce these thermal effects, 0.5 to 15 parts per hundred (pph) of a stabilizer is mixed in the dispersion composition. Effective stabilizers include metallic soaps, organic tin and cadmium salts, and epoxide resins.

Vinyl and vinylidene fluoride systems, although more thermally stable than their vinyl chloride cousins, undergo thermal decomposition at relatively high temperatures (>600°F). The process is greatly accelerated in the presence of glass or silica and these materials are to be avoided in formulating these systems. Copper, aluminum, and iron show no catalytic degradation effects; although, surprisingly, titanium dioxide shows a tendency to discolor the vinylidene fluoride systems and discoloration becomes more pronounced with increasing pigment concentrations. Effective stabilizer products for these dispersion resins are calcium-zinc complexes or pentaerythritol in combination with an antioxidant.

It has been found that film integrity, gloss, flexibility, and process or hot water resistance are materially affected by the baking cycles used. In practice, the fusing cycle must be especially established for each application. In general, optimum results are obtained when the coated metal is exposed to relatively high temperatures in the first oven zone. As a result, the solvents in the dispersion composition have an opportunity to help solvate the dispersion resin before they evaporate. Of course, care must be taken to adjust the temperatures in the first oven zones so that no blistering or pinholes occur. Based on practical and theoretical consideration with PVC dispersion coatings, it was determined that 335°F (169°C) is the minimum metal temperature at which the PVC dispersion should be fused for good results. A recommended sheet-coating schedule, for example, would be 340° to 365°F (171° to 185°C) for 6 to 10 minutes. A representative coil bake would be 60 seconds at 500°F (260°C) for 0.6 mil film. Properly cured films thus baked achieve tensile strengths of 8000 to 10000 psi (55 to 69 MPa). Elongation is approximately 300%. Similar curing considerations apply to the fluoropolymers. However, these coatings require considerably more energy to properly coalesce the film. Metal temperatures of 475°F (245°C), approximately 100°F (38°C) above those used for the vinyl chloride systems, are required. A representative coil bake for these resins, for example, would be 425° to 475°F for 30 to 45 seconds.

NONFOAM STRIPPABLE VINYL

Another group of chemical coatings, the uses of which have shown continued marked expansion, are the nonfoam strippable vinyls. While these materials have been offered for some time, they were formulated for spray application to products after fabrication. The more recent types, like the roll-coat finishes, are designed for application by reverse roller coating to coiled metal before the product is manufactured. Therefore, they offer surface protection all the way through metalworking operations, during assembly, and many times afterward as preliminary packaging.

These types generally consist of vinyl plastisols applied in liquid form and heat-converted into a continuous film, generally at a minimum of about 2 mils dry. Here again improved resins have played an important part in the superior performance of these materials by providing the materials with excellent toughness as well as tensile and tear strength to withstand slitting, stamping, forming, and bending.

Formulated with just the right degree of cohesive properties to adhere until no longer desired, these strippables can be used over a variety of substrates including polished or stainless steel, anodized aluminum, or prefinished metal that has been coated with TS finishes.

Converters report different applications of their plastisol strippable vinyl in which users reduced material and labor costs 50% by adopting this concept. The firms use the strippable vinyl on anodized aluminum coil that is subsequently manufactured into products such as heating hoods. For this application the strippable vinyl remains intact before, during, and after fabrication; acts as preliminary packaging and protection against scratching from the final corrugated container; and stays on until the hood is installed to protect it from installation handling.

Prior to using the plastisol strippable, companies that produce heating hoods employed pressure-sensitive paper. This material was almost twice the cost per square foot of plastisol strippables and had to be removed before the hood was shipped; thus additional packaging had to be used.

Architectural firms also use strippable plastisol coatings on stainless steel building panels. Here they offer surface protection from, before, and during fabrication of the panels, up to the time at which they are erected.

FOAM-VINYL STRIPPABLE

The foam-vinyl strippables are very useful for packaging metal parts and other similar products. Based on PVC dispersion resins, foam vinyl strippables are applied in liquid form to the completed product. Foaming takes place during their cure cycle to produce a highly resilient, spongy film. Therefore, these strippables also offer protection against denting and scratching, and have taken the place of paper and corrugated wrappings at substantial savings.

Within the past few decades, types of strippables that can expand up to 300% have been made available. These types can yield maximum films of about ½ in, although films of ¼ in are more commonly used. These materials afford many advantages. For instance, they retard corrosion by forming a tight skin around the object, which inhibits the entrance of moisture. They also help to save space since this tight fit allows more units to be stacked per cubic foot than if bulky containers were used. In addition, because one type of strippable can accommodate products of all sizes and shapes, there is no need to maintain a large inventory of different-sized packaging materials. While auto parts packaging is one of their most common uses, foam vinyl strippables are also used in other industries in which metal parts shipment prevails.

PLASTICIZER, PVC

This review of plasticizers is required because they play an important part in the environmental performance of vinyl. Plasticizers serve three basic functions: to lower the processing temperature of the polymer below the decomposition temperature, to modify the properties of the finished product, and to modify the processing properties (chapter 1). Plastics can be made softer or flexible, their natural properties can be extended or modified, and their viscosities can be reduced to make them easier to shape and form at high temperatures and pressures.

The mechanism by which inclusion of plasticizers in PVC results in increased flexibility is attributed to a reduction of the intermolecular forces. In other words, the plasticizer acts as a lubricating agent to allow the macromolecules to slide over one another freely, or by the solvation of the polymer. Unplasticized PVC is a three-dimensional gel formed by the attachments between molecules at active centers. Plasticization is a reduction of polymer-polymer unions by creating polymer-plasticizer unions in their place.

Plasticizers for PVC are generally divided into two categories: true plasticizers (primary and secondary) and extenders. The primary types are materials that exhibit good compatibility with PVC. The secondary types usually exhibit fair to good compatibility and are normally used in conjunction with the primary plasticizers. One highly desirable property of a plasticizer is its capacity to impart and maintain the characteristics of an elastomer to the polymer over the widest possible temperature range. Unfortunately, no single plasticizer embodies all the desired combinations of properties. Therefore, for any specific application, it is necessary to choose the correct plasticizer combination.

FLUORINE-CONTAINING RESIN

The high thermal stability of the carbon-fluorine bond has led to considerable interest in fluorine-containing polymers as heat resistant plastic and rubbers such as polytetrafluoroethylene (PTFE). PTFE is a material that also provides exceptional chemical resistance. It is completely inert to halogens, fuming mineral acids, strong alkalis, and oxidizing agents. It also has the advantage of being nonflammable. However, it is attacked by molten alkali metals. Its insolubility in organic solvents makes it impossible for use in lacquers. With its high crystalline melting point of 330°F (166°C), it cannot be used in organosols and plastisols. Other fluorine-containing plastics have been developed, but in almost all cases they are not available for use in the coating industry.

ACRYLIC RESIN

Acrylic resins are TP polymers ranging from very hard and tough to extremely flexible water-white materials. They are resistant to oxidation, ultraviolet degradation, and many chemicals. However, certain solvents soften them.

They have been used for many years in specialty coatings. Acrylic resins have established a reputation for excellent durability in automotive lacquers. They can be used with plasticizers as the film former or in combination with nitrocellulose. The combination is somewhat harder and has better resistance to solvents such as gasoline.

The very flexible types of acrylic resin have been used as coatings for elastic materials, such as rubber, and for textile or leather coatings. Acrylic resins are compatible with many other film formers and are often used in blends. Acrylic emulsions are used as binders in latex paints; the latex paints have steadily increased in use as architectural coatings. Applications include interior plaster and exterior stucco, concrete, and masonry surfaces.

The disadvantages that stem from the TP nature of the ordinary acrylic resins (solvent sensitivity and temperature sensitivity) have been overcome by the introduction of thermosetting acrylic resins. They are cross-linked by stoving processes, very similar to those used for the alkyd-melamine types. Water-soluble or emulsion types, as well as those soluble in organic solvents, are all available.

CELLULOSIC RESIN

Nitrocellulose was the first synthetic high polymer used in coatings. Its lacquers are still considered to be the fastest air-drying materials. This is due largely to its high softening point and good solvent release. It is compatible with many other resins and plasticizing materials.

It provides hard furniture finishes, flexible coatings for paper and fabrics, and durable finishes for automobiles. The principal limitations of such lacquers are their relatively high-cost solvents and relatively low solids at spraying consistency, their sensitivity to heat and ultraviolet radiation, and their high degree of flammability.

Ethyl cellulose is softer and more flexible than nitrocellulose. It is not as highly flammable as nitrocellulose and has a certain degree of resistance to degradation by heat. Chemical resistance is improved. They provide toughness in blends with hard resins and waxes for hot-melt coatings.

Cellulose acetate is used chiefly in plastics and sheeting. It has only a few specialty applications in coatings. However, the acetate-butyrate has increasing uses. Its properties vary with the ratio of acetyl and butyral groups. It is slightly softer than ethyl cellulose but has better resistance to moisture absorption.

COPOLYMER RESIN

Many different copolymers are used as binders in surface coatings. A few of these types have been reviewed in this chapter (chapter 2). The styrene-butadiene copolymers are very popular. Many different vinyl copolymers—such as vinyl chloride, vinyl isobutyl ether, vinylidene chloride, vinyl acrylonitrile, and so on—are also used.

COUMARONE-LNDENE RESIN

The coumarone resins are materials of low MW. They are soluble in aliphatic, aromatic hydrocarbons and in oils. They are TPs, and because of their hydrocarbon character they are resistant to acid and alkalis.

They show a marked tendency to yellow when exposed to light, and their durability is poor. Because of these inherent restrictions, they have limited applications in coatings. Their main field of use is in binders for aluminum and bronze lacquers, where their low acid value leads to stability in the ready-mixed finish. As they are protected from the atmosphere by the layers of bronze pigment, the durability of such finishes is surprisingly good. Improvements occur by adding tung oil. When added in asphalt lacquers, improved gloss and alkali resistance occurs.

PARYLENE

This plastic permits pinhole-free coatings with the exceptional capability of producing outstanding conformity and thickness uniformity. Ultrathin (35 to 3000 nm) parylene films are produced called pellicles are produced. These coatings are used to protect units from airborne contaminants, moisture, salt spray, and corrosive vapors while maintaining excellent insulator protection. The coatings are also extensively used in the protection of hybrid circuits. Such coatings do not affect part dimensions, shapes, or magnetic properties.

These transparent TPs are generally insoluble up to 302°F (150°C). At 518°F (270°C) they will dissolve in chlorinated biphenyls, but the solution gels upon cooling below 320°F (160°C). Their weather resistance is poor. Embrittlement is the primary consequence of their exposure to ultraviolet radiation.

Parylene has a unique coating system. Also unique is the chemistry of the xylene monomer, in which a substrate is exposed to a controlled atmosphere of pure gaseous monomer, *p*-xylylene

(PX). The coating process is best described as a vapor deposition polymerization (VDP). The monomer itself is thermally stable but kinetically unstable. Although it is stable as a gas at low pressure, upon condensation it spontaneously polymerizes to produce a coating of a high-molecular-weight, linear poly (*p*-xylylene; PPX).

The *p*-xylylene polymers (PPXS) formed by the Gorham process are generically known as the parylenes. The terms Parylene N, Parylene C, Parylene D, or Parylene HT refer specifically to coatings produced from the original patents from Union Carbide Corporation's dimers. The polymerization process takes place in two stages that must be physically separate but temporally adjacent.

There are four primary variants of the polymer: Parylenes N, C, D, and HT. Although they all have the same essential coating properties and are applied in the same manner, each has a unique molecular form that results in specialized performance characteristics. Parylenes N and C are the most commonly used variants; they are used in medical coating applications. Table 10.14 describes the key properties of these parylenes (478).

Of all the variants, Parylene N offers the highest penetrating power. Because of its greater molecular activity in the monomer phase, it can be used to coat relatively deep recesses and blind holes. This form of parylene also provides slightly higher dielectric strength than Parylene C and a dielectric constant that is independent of frequency. The lower dissipation factor and dielectric constant of this parylene form enable it to be used for protecting high-frequency substrates in which the coating is in the direct electromagnetic field.

Parylene C differs from Parylene N in that it has a chlorine atom on the benzene ring, providing a useful combination of electrical and physical properties. Among these properties is a very low permeability to moisture and corrosive gases. Compared to Parylene N, Parylene C displays weaker crevice-penetrating ability.

The parylene process has certain similarities to vacuum metallizing. The principal distinction is that truly conformal parylene coatings are deposited even on complex, 3-D substrates, including on sharp points and into hidden or recessed areas. Vacuum metallizing, on the other hand, is a line-of-sight coating technology. Areas of the substrate that cannot be seen by the evaporation source are shadowed and remain uncoated.

Parylenes have been used as beam splitters in optical instruments, windows for nuclear radiation–measuring devices, dielectric supports for planar capacitors, dielectric film in high-performance precision electrical capacitors, circuit boards, and electronic module coatings. They have also been used for extremely fast-responding, low-mass thermistors and thermocouples, and medical devices. Parylene has been used in a wide range of medical applications since the 1970s. These include catheters and mandrels, stents, needles, cannulae, cardiac-assist devices, prosthetics, and circuitry. Certain devices require a protective coating to isolate them from contact with moisture, gases, corrosive biofluids, or chemicals. The different parylenes respond to sterilization in different ways, as shown in Table 10.15. Coatings are also used to protect patients from contact with surgical items or implanted devices that may not be biocompatible. Vacuum-deposited parylene is often the protective medical coating of choice. Additionally, parylene may be used to deliver other functional properties, such as electrical insulation, particulate tie-down, or increased lubricity (478).

Property		Parylene N	Parylene C
Dielectric constant	60 Hz	2.65	3.15
	1 KHz	2.65	3.10
	1 MHz	2.65	2.95
Dissipation factor	60 Hz	0.0002	0.020
	1 KHz	0.0002	0.019
	1 MHz	0.0006	0.013
Secant modulus (psi)		350,000	400,000
Tensile strength (psi)		6000–11,000	10,000
Yield strength (psi)		6100	8000
Elongation to break (%)		20–250	200
Yield elongation (%)		2.5	2.9
Density (gm/cm³)		1.10–1.12	1.289
Index of refraction (n_D^{23})		1.661	1.639
Water absorption (% after 24 hr)		<0.1	<0.1
Rockwell hardness		R85	R80
Static coefficient of friction		0.25	0.29
Dynamic coefficient of friction		0.25	0.29
Melting point (°C)		420	290
T5 point (°C)		160	125
Gas permeability at 25°C (cm³(STP)•mil/100 in²/d•atm)	N_2	7.7	1.0
	O_2	39	7.2
	CO_2	214	7.7
	H	540	110
Moisture vapor transmission at 90% RH, 37°C (g• mil/100 in²• d)		1.5	0.21

Table 10.14 Examples of properties for Parylenes N and C

Sterilization Method	Parylene N					Parylene C				
	Dielectric Strength	MVT	Tensile Strength	Tensile Modulus	COF	Dielectric Strength	MVT	Tensile Strength	Tensile Modulus	COF
Steam	None	Δ43%	None	Δ12%	Δ38%	None	Δ5%*	Δ17%	Δ9%	None
EtO	None	Δ21%	None	None	Δ33%	None	Δ8%	None	None	None
E-beam	na	None	None	None	None	NA	None	None	None	None
H_2O_2 plasma	None	None	None	None	Δ48%	Δ9%	None	None	None	Δ188%
Gamma	None	None	None	None	None	None	Δ5%*	None	None	None

*5% values ares not likely to be statistically significant. NA=not applicable.

Table 10.15 Effect of various sterilization methods for Parylenes N and C

Applying parylene requires special, though not complex or bulky, equipment: a vaporizer, a pyrolysis unit, and a deposition chamber. The objects to be coated are placed in the deposition chamber, where the vapor coats them with a polymer. A condensation coating like this does not run off or sag as in conventional coating methods, nor is it line-of-sight technology, as in vacuum metallizing. In condensation coating, the vapor evenly coats edges, points, and internal areas. Although the vapor is all-pervasive, holes can still be coated without bridging. Masking can easily prevent chosen areas from being coated. The objects to be coated can also remain at or near room temperature, thus preventing possible thermal damage. The quantitative nature of this reaction allows the coating thickness to be accurately and simply controlled by manipulating the polymer composition charged to the vaporizer.

PROCESS

The coating process technology of applying parylene film to a substrate involves a vacuum chamber by means of VDP. A dry, powdered precursor known as a dimer is converted by heat in the coating system to form a dimeric gas, and heated further to generate a monomer gas that is passed to a deposition chamber.

Within the chamber, it polymerizes at room temperature as a conformal film on all exposed substrate surfaces. Parylene deposition has no liquid phase, uses no solvent or catalyst, and generates no gaseous by-products. There are no cure-related hydraulic or liquid surface-tension forces in the coating cycle, and coated products remain free of mechanical stress.

This film becomes a linear, crystalline polymer with an all-carbon backbone and a high MW. With the absence of polar entities, and substantial crystallinity, the film is stable and highly resistant to chemical attack. The static and dynamic coefficients of friction for parylenes are in the range of 0.25 to 0.33. This dry-film lubricity is an important characteristic for certain applications, such as medical catheter and guide-wire coatings.

APPLICATION

Coated products are all around us worldwide. This large industry produces two broad categories of coatings, namely, the trade sales and the industrial finishes. Trade sales, or shelf goods, include products sold directly to consumers, contractors, and professional painters for use in construction or painting, refinishing, and general maintenance. These coatings are used chiefly on houses and buildings, although a sizeable portion is used for refinishing automobiles and machinery.

Industrial finishes, or chemical coatings, encompass myriad products for application by manufacturers in factories or for industrial maintenance and protection. They are custom-made products sold to other manufacturers for such items as automobiles, appliances, furniture, ships and boats, metal containers, streets and highways, and government facilities.

Different applications are reviewed in this chapter. The following just provide additional applications. Table 10.16 provide a guide on painting plastics, where R = recommended and NR = not recommended. To improve or provide bonding capabilities to NR substances, different primers or surface treatments or both are used such as fluorination (Table 10.17; chapter 6).

Plastic	*Urethane*	*Epoxy*	*Polyester*	*Acrylic lacquer*	*Acrylic enamel*	*Acrylic waterborne*
ABS	R	R	NR	R	R	R
Acrylic	NR	NR	NR	R	R	R
PVC	NR	NR	NR	R	R	NR
Styrene	R	R	NR	R	R	R
PPO/PPE	R	R	R	R	R	R
Polycarbonate	R	R	R	R	R	R
Nylon	R	R	R	NR	NR	NR
Polypropylene	R	R	R	NR	NR	NR
Polyethylene	R	R	R	NR	NR	NR
Polyester	R	R	R	NR	NR	NR
RIM	R	NR	NR	NR	R	R

Table 10.16 Guide for applying paint coatings to plastic substrates

Plastic	Surface energy	
	before mN/m	after mN/m
Polystyrene	35	72
EPDM	40	58
Polycarbonate	35	72
Polyacetal	40	72
Polyphenylene sulfide	45	58
Polyvinylidene fluoride	35	54
Polyimide	40	58
Polyetherimide	35	56
Polyethersulfone	40	72
Polysiloxane	32	58

Table 10.17 Surface energy of plastics as a result of fluorination

COIL COATING

This section reviews coil coating and highlights a typical major application of coil coating. Coil coating with plastics continues to be a very big business worldwide (Tables 10.18 and 10.19). Table 10.20 reviews the properties of coil coating plastics. In the meantime, the coil coating industry has been under pressure to eliminate the use of solvents. For example, in the past, TP polyester

Material	Main uses	Methods of application	Average coating thickness (in)
Low density polythene	Decoration and protection of domestic articles such as wirework, clips, hooks and clamps	Fluidized bed, flock spraying, flame spraying	0·040
High density polythene	Corrosion resistant coatings in industrial applications, metal pipe fittings, tool handles, electroplating jigs, switch covers	Fluidized bed, flock spraying, spraying in dispersion form, flame spraying	0·030
Nylons 11 and 12	Coating of metal furniture, pipework, domestic and electrical appliances	Fluidized bed, electrostatic spraying, flock spraying, spraying in dispersion form, flame spraying	0·020
Epoxide resin	General coating of irregular shaped articles, electrical components and internal coatings of steel pipes	Fluidized bed, electrostatic spraying	0·005
Chlorinated polyether	Versatile corrosion resistant lining and coating of chemical plant	Fluidized bed, flock spraying, flame spraying	0·020
PTFE	Baking tins, household pans, industrial molds as a release and anti-stick surface	Spraying in dispersion form	0·001
PTFCE	Corrosion resistant and electrical insulation applications	Spraying in dispersion form	0·008
Polyurethane	Vibratory feed hoppers, pulley wheels, linings for shot blasting equipment where excellent abrasion resistance is required	Fluidized bed, flock spraying	0·030
Plasticized PVC	Protection of industrial pipework and plant against corrosion and abrasion. Cushion coating for racks and stillages and for low temperature flexibility	Plastisol dipping, fluidized bed, spraying in dispersion form	0·020–0·125

Table 10.18 Typical plastics used in coil coatings

Coatings	Characteristics	Applications
Vinyls	Excellent flexibility and exterior durability; can withstand rigorous fabrication	Aluminum screen doors, TV cabinets, dashboard panels, exterior siding
Alkyds	Good exterior durability and color retention; economical; not suitable for severe forming operations	Exterior finishes for metal containers, drapery hardware, shelving, any decorative purposes not involving severe forming
Plastisols, organosols	Excellent scuff and mar resistance, color durability; good formability; both belong to vinyl dispersion family, but plastisols (100% solids) yield thicker finishes; both require primer for adhesion	TV cabinets, metal furniture, gasketing, interior wall paneling
Epoxies	Excellent hardness, toughness, flexibility, abrasion resistance; not recommended for exterior use	Sanitary coating, ash trays, cap coating
Phenolics	Good stain, acid, solvent resistance; poor flexibility and adhesion; require modified vinyl size coat to withstand normal bends	Beer can interior lining, closures
One-coat organosols	Excellent forming, durability properties; high solids content; similar to conventional vinyl solution coatings	Roof decking, exterior siding, cabinets
Vinyl–alkyds	Compromise in price and properties between vinyl and alkyds; good exterior durability	Exterior siding, applications where straight alkyd would be used but where more than normal fabrication is involved
Acrylics	Excellent stain, abrasion, mar resistance; good durability; limited formability, full-gloss appearance, acceptable for decorative uses	Appliance housing, cabinets, lighting fixtures, exterior siding, refrigerators, plastisol overfinishes

Table 10.19 Coil coating plastic characteristics and applications

coil coatings contained up to 40% of solvents such as glycol esters, aromatic hydrocarbons, alcohols, ketones, and butyl glycol. It has been predicted that the solvent-based technology will not change during the next decade because the industry heavily invested in equipment to handle solvents (374).

Such changes in technology require long testing before they can be implemented. The coil coating industry normally recovers energy from evaporated solvents either by at-source incineration or by a recycling process that lowers emissions. Because of the large amount of solvents used, the use of PVC and fluoropolymers in some formulations, and the use of chromates in pretreatments, pressure remains on the industry to make improvements. The coil coating industry is estimated consume about 50000 tons of solvents both in Europe and in the United States. About half of these solvents are hydrocarbons.

According to the published studies, efforts to change this situation started in the early 1990s. By the mid-nineties research data were available to show that the technology can be changed. Two directions that will most likely challenge the current technology are radiation curing and powder coating.

Coil coats are thin (about 30 μm wet thickness) but contain a high pigment loading. Consequently, ultraviolet curing is less suitable than electron beam curing. The application of this technology requires a change to the plastic system, and acrylic oligomers are the most suitable for this application. These systems can be processed without solvents. If a reduction in viscosity is required, it can be accomplished with plasticizers (the best candidates to date are branched phthalates and linear adipates) or reactive diluents, such as multifunctional monomers, or both. Results show that the ultraviolet stability of the system needs to be improved by using a polyester topcoat or fluoroplastic.

Performance rating 1—Excellent 2—Good 3—Fair 4—Poor	Thin films, 0.1–1.2 mils										Thick films		Laminates	
	Amine-alkyd	Vinyl-alkyd	Thermoset acrylic	Solution-vinyl	Oil-free polyester	Epoxy-ester	Straight epoxy	Silicone alkyd	Polyvinyl fluoride	Polyvinylidene fluoride	Organosol	Plastisol	Polyvinyl fluoride	Polyvinyl chloride
Appearance considerations														
Ability to achieve high gloss (above 85 units 60°)	1	2	1	1	1	2	1	1	3	4	4	4	4	4
Resistance to color fading	3	3	2	2	2	4	4	1	1	1	2	2	1	2
Long-time retention of color and gloss	3	3	1	3	1	3	3	1	2	2	3	3	2	3
Fabricating properties														
Film adhesion	2	2	1	1	2	1	1	2	2	2	1	1	2	1
Film flexibility	3	2	2	1	2	3	3	3	1	1	1	1	1	1
Adaptability to embossing of substrate	4	3	3	2	3	3	4	4	1	1	1	1	1	1
Resistance to metal marking	2	2	2	2	2	2	1	2	3	3	3	3	2	3
Ability to fabricate after aging in storage	4	2	2	1	2	3	3	3	1	1	1	1	1	1
Performance in service														
Film hardness	2	2	2	2	1	2	1	2	2	2	2	3	2	3
Abrasion resistance	3	2	2	2	2	2	2	2	1	1	1	1	1	1
Mar resistance (fingernail test)	2	2	1	2	1	2	1	2	3	3	3	3	2	3
Stain resistance (food and household agents)	3	3	2	2	2	2	1	2	1	1	2	2	1	2
Resistance to grease and oil	2	2	1	1	2	2	1	2	1	1	1	1	1	1
Outdoor exposure														
General corrosion resistance, industrial atmospheres	2	2	2	2	2	2	1	2	1	1	1	1	1	1
Salt-spray resistance	3	2	1	1	2	1	1	2	1	1	1	1	1	1
Weather durability, pigmented	2	3	2	2	2	4	4	1	1	1	1	1	1	1
Weather durability, clear films	3	3	2	4	3	4	4	2	1	1	4	4	3	4
Resistance to solvents and chemicals														
General chemical resistance—acids and alkalis	3	3	2	2	3	2	1	2	1	1	1	1	1	1
Resistance to aliphatic hydrocarbon solvents	2	2	1	2	2	1	1	2	1	1	1	1	1	1
Resistance to aromatic hydrocarbon solvents	2	3	2	4	2	1	1	2	1	1	3	3	1	3
Resistance to ketones or oxygenated solvents	3	3	2	4	3	2	1	2	1	1	4	4	1	4

Table 10.20 Plastic properties of coil coatings

With a topcoat, the materials perform very well, as observed in laboratory experiments and in industrial environments. At the time of the study, which took place in the mid-nineties, the process of coating was less efficient than solvent-based systems because production speed was about six times slower than the highest production rates in the industry (120 m/min). At the same time, it is known that the quality of solvent-based coatings suffers from excessive production rates. Radiation curing has a disadvantage because of its high capital investment, but it does have an economics advantage because the process is very energy efficient. Radiation curing technology has been successfully implemented in several industries, such as paper, plastic processing, and wood coating, where long-term economic gains made the cost viable.

Comparisons of solvent-based fluoroplastic powder coating developed in Japan show that the elimination of solvent is not only good for the environment but also improves performance (ultraviolet stability especially is improved). The study was carried out with a very well-designed testing program to evaluate the weathering performance of the material.

These two technologies show that there is extensive activity to improve coil coatings with simultaneous elimination of solvents. Two recent patents contribute more information on the developments in the coil coating industry. One problem in the industry is the poor adhesion of the coating to steel. There is a primer that contains dipropylene glycol methyl ether and PM acetate that allows the deposition of relatively thick layers (20 to 40 μm) without blistering and at suitable rate of processing. However, the primer has a low solids content (30% to 45%). A new retroreflective coating that is based on ethyl acrylate-styrene copolymer was developed, which contains a mixture of xylene with another aromatic hydrocarbon at relatively low concentration (11% to 12%).

STRIPPABLE COATING

There are different plastic types of strippable coatings to meet different requirements. A popular type uses vinyls that are for protecting metal parts being packaged for shipment. PVC dispersion plastic is applied in liquid form to the product. Foaming takes place during their cure cycle to produce a highly resilient, spongy film. Therefore, these strippables also offer protection against denting and scratching, and have taking the place of paper and corrugated wrappings at substantial savings. Types that can expand up to 300% have been made available. These types can yield maximum films of about ½ in, although ¼ in is more commonly used.

Spraying, dipping, flow and curtain coating can apply foam vinyl strippables over the same substrates as the nonfoam types (Fig. 10.3). Their primary use is on chrome-plated automotive replacement parts, such as bumpers, headlight bezels, and decorative trim. Strippables usually

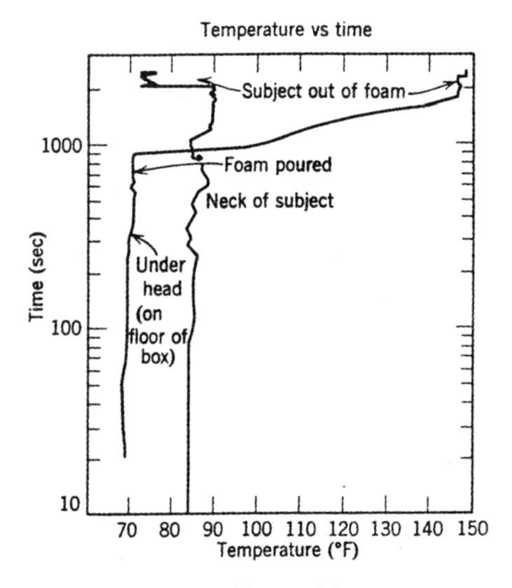

Figure 10.3 Temperature distribution in strippable vinyl foam.

boost production of wrapped parts considerably, as well as lowering the reject rate. Output of wrapped bumpers, for example, was increased by three times.

GERM-FREE COATING

Past attempts to create surfaces with inherent bactericidal properties capable of rendering them free of germs have been unsuccessful. Researchers at Northeastern University, working with colleagues at the Massachusetts Institute of Technology and Tufts University (all in the Boston area), believe they may have developed a method for creating permanently germ-free dry surfaces (479). They speculate that previous efforts to design dry bactericidal surfaces failed because the polymer chains that made up the material were not sufficiently long and flexible enough to penetrate bacterial cell walls.

Their research has demonstrated that covalent attachment of N-alkylated poly(4-vinylpyridine; PVP) to glass can make surfaces permanently lethal to several types of bacteria on contact. The group found a narrow range of N-alkylated PVP compositions that enable the polymer to retain its bacteria-killing ability when coated on dry surfaces. It is believed that these are the first engineered surfaces proven to kill airborne microbes in the absence of a liquid medium.

Work previously conducted on different compositions had limitations. Their polymer chains had insufficient length and flexibility. Their polymer includes a long linker that enables the toxic N-alkylated pyridine groups to cross the bacterial envelope.

According to the Boston-area researchers cited previously, dry surface-bonded PVP with no N-alkyl chains or long N-alkyl chains, including ten or more carbon units, is not bactericidal. They reported that three- to eight-unit PVP chains derive a sufficient positive charge from the cationic pyridine nitrogen to repel each other and stay flexible and sufficiently hydrophobic to penetrate bacterial cell walls. It has been indicated that surfaces fabricated in this way kill 94% to 99% of bacteria sprayed on them. Because the coating is chemically bonded to the surface, it will not be affected by being touched or washed.

EVALUATION METHOD

Severe near-future requirements for resin protective coatings demand the use of all available methods of characterizing candidate materials. Examples include thermal, optical, and electrical methods. The two main thermal methods to consider are thermogravimetric analysis (TGA) and differential thermal analysis (DTA; chapter 22). Both may be used to characterize potential coating materials under conditions that would provide information for the best selection, formulation, and application of these materials by investigating their thermal degradation patterns and mechanisms.

The optical methods of interest are spectrophotometric and photomicrographic. Spectrophotometry is used to investigate the changes in optical properties of coatings that have been subjected to various environmental conditions. Photomicrography can be used to either examine or determine the metal-coating interface. It can also be used to determine if a coating is crystalline, amorphous, continuous, or lacking in integrity.

An important electrical method is the measurement of the dielectric breakdown point of a coating. The instruments that are used for this purpose can also be used to determine the porosity and uniformity of a coating (chapter 22).

PROCESS

OVERVIEW

Both TPs and TS plastics may be used as coatings. The materials to be coated may be plastic, metal, wood, paper, fabric, leather, glass, concrete, ceramics, and so on. Methods of coating are varied, as shown in Table 10.21:

Coating method	Base material[a]	Coating composition[b]	Usual coating speed ($m\ min^{-1}$)	Viscosity range, (m Pa s)	Wet-coating thickness range (μm)
Air knife	B,D	R,T,X	15–600	1-500	25–60
Brush	A,B,C,E,F,G	R,S,X,Z	30–120	100–2,000	50–200
Calender	A,B,D,E	U,V,W	5–90		100–500
Cast-coating	A,B,D	Q,R,S,T,V,Y	3–60	1,000–5,000	50–500
Curtain	A,B,C,D,E,F	R,S,V,X,Z	20–400	100–20,000	25–250
Dip	A,B,D,E,F,G	R,S,V,X,Y,Z	15–200	100–1,000	25–250
Extrusion	A,B,D,E	T,U,V,W	20–900	30,000–50,000	12–50
Blade	A,B	R,S,T,V,X,Y,Z	300–600	5,000–10,000	12–25
Floating knife	A,B,D	R,S,T,V,X,Y,Z	3–30	500–5,000	50–250
Gravure	A,B,D,E	R,S,T,U,V,Y,Z	2–450	100–1,000	12–50
Kiss roll	A,B,C,D,E,F	R,S,V,X,Z	30–300	100–2,000	25–125
Knife-over-blanket	A,B,D	R,S,T,V,X,Y,Z	3–30	500–5,000	50–250
Knife-over-roll	A,B,C,D,E	R,S,T,U,V,X,Y,Z	3–60	1,000–10,000	50–500
Offset gravure	B,D	R,S,T,Z	30–600	50–500	12-25
Reverse roll	A,B,C,D,E,F	R,S,T,U,V,X,Y,Z	30–300	50–20,000	50–500
Reverse-smoothing roll	A,B	R,T,X	15–300	1,000–5,000	25–75
Rod	B,D	R,S,T,V,X,Y,Z	3–150	50–500	25–125
Sprays					
Airless spray	A,B,C,D,E,F,G	S,T,V,X,Y,Z	3–90	-	2–250
Air spray	A,B,C,D,E,F,G	S,T,V,X,Y,Z	3–90	-	2–250
Electrostatic	A,B,C,D,E,F,G	S,T,V,X,Y,Z	3–90	-	2–250
Squeeze roll	A,B,C,D,E,F	R,S,T,U,V,X,Y	30–700	100–5,000	25–125
In situ polymerization	A,B,C,D,E,F,G	Y,Z	undetermined	liquid or vapor	6–2.5
Powdered resin	A,B,C,E,F,G	Q	3–60		25–250[c]
Electrostatic spray		Q			20–75[c]
Fluidized bed	E,G	Q			200–2,000[c]

Table 10.21 Coating methods related to performances

BASE MATERIAL

A = woven and nonwoven fabric
B = paper and paperboard
C = plywood and pressed fiberboard
D = plastic films
E = metal sheet, strip, or foil
F = irregular flat products
G = irregularly shaped products

COATING COMPOSITION

Q = powdered resin compositions
R = aqueous latexes, emulsions, dispersions
S = organic lacquer solutions and dispersions
T = plastisol and organosol formulations
U = natural and synthetic rubber compositions
V = hot-melt compositions
W = TP masses
X = oleoresinous composition
Y = reacting formulations (e.g., epoxy and polyester)
Z = plastic monomers

The processes include extrusion (Fig. 10.4; chapter 5); roller coating (Fig. 10.5); knifing or spreading (Fig. 10.6); transferring (Fig. 10.7); cast-transferring (Fig. 10.8); dipping (Fig. 10.9); vacuuming (Fig. 10.10); in-mold via reaction injection molding (Fig. 10.11; chapter 12); electrodeposition (Fig. 10.12); spraying (Table 10.22); fluidized bed; brushing; floccing; microcapsulation; radiation; and many others (a few will be reviewed). Calendering of a film to a supporting material is also a form of coating that tends to be similar to roll coating (chapter 9). Processes are also used to coat specific products such as floor covering (Fig. 10.13) and foamed carpet backing (Fig. 10.14).

Surface coatings are usually composed of viscous liquids. They have the three basic components of a film-forming substance or combination of substances: a binder, a pigment or combination of pigments, and a volatile liquid. The combination of binder and volatile liquid is usually called "the vehicle." It may be a solution or a dispersion of fine binder particles in a nonsolvent. No pigments are included if a clear, transparent coating is required. The composition of the volatile liquid provides enough viscosity for packaging and application, but the liquid itself rarely becomes part of the coating.

Film coatings can involve chemical reactions, polymerization, or cross-linking. Some films merely involve coalescence of plastic particles. The various mechanisms involved in the formation of plastic coatings are as follows:

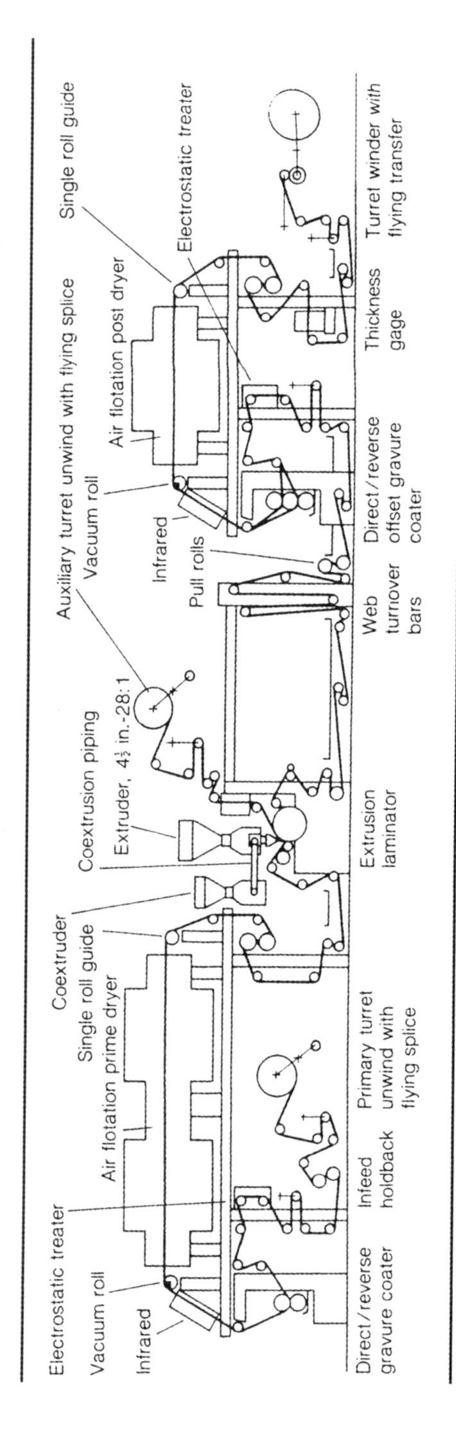

Figure 10.4 High-speed extrusion coating line.

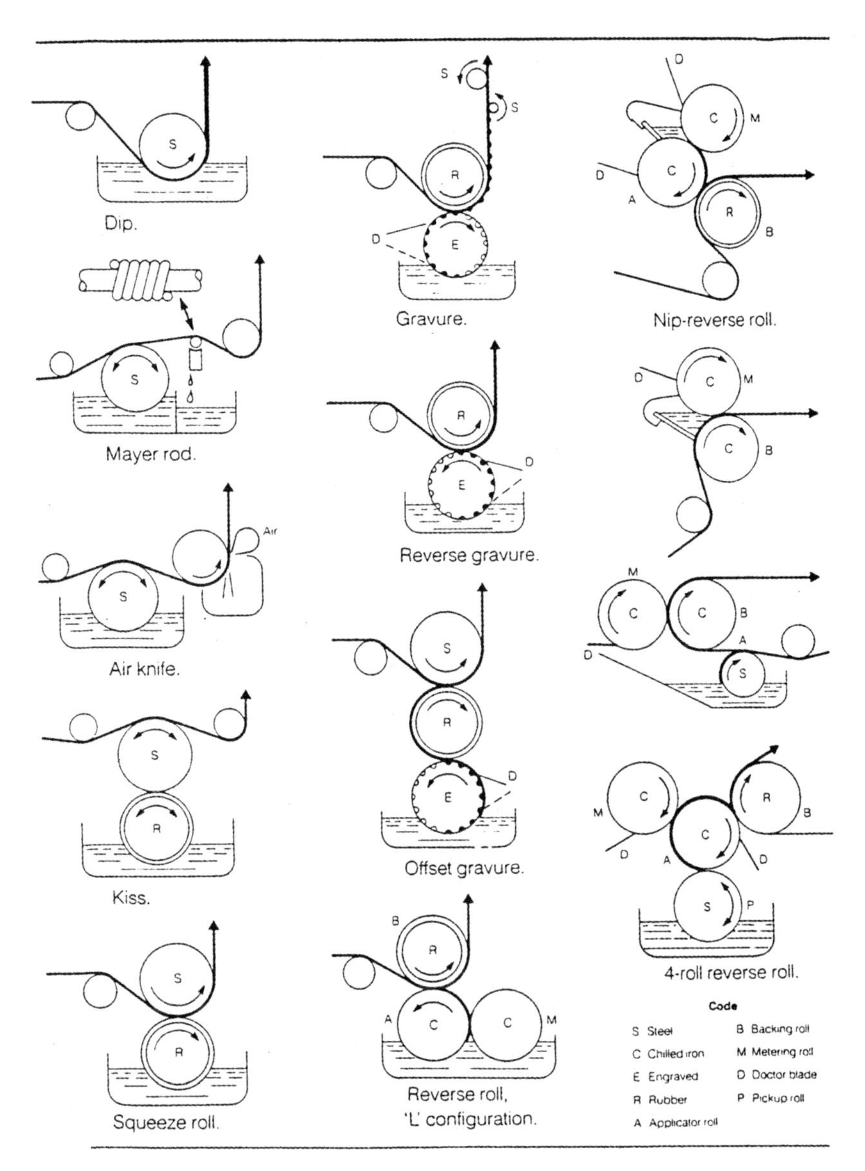

Figure 10.5 Example of roller coating processes.

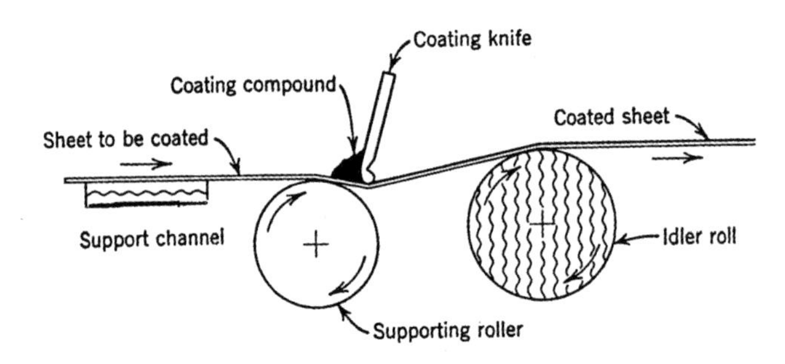

Figure 10.6 Knife spread coating.

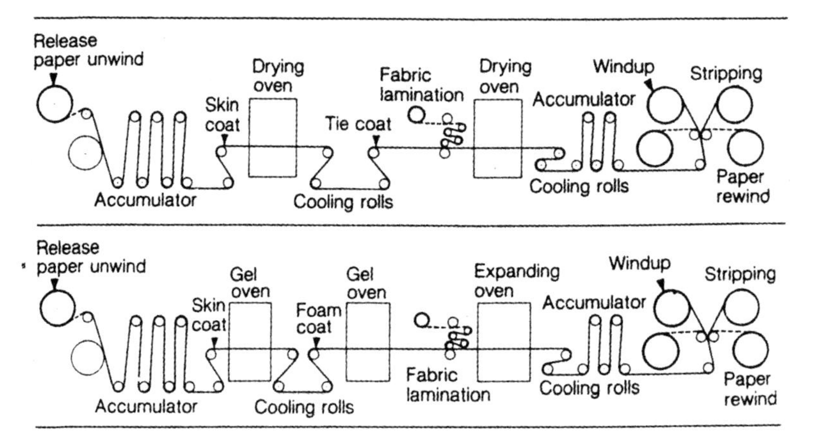

Figure 10.7 Transfer coating of PUR (top) and PVC.

1. Coating formed by chemical reaction, polymerization, or cross-linking of epoxy, TS polyester, PUR, phenolic, urea, silicone, and so on
2. Dispersions of a plastic in a vehicle; after removal of the vehicle by evaporation or bake, the plastic coalesces to form a film of plastisol, organosol, water-based or latex paint, fluorocarbons, and so on
3. Plastic dissolved in a solvent followed by solvent evaporation to leave a plastic film of vinyl lacquer, acrylic lacquer, alkyd, chlorinated rubber, cellulose lacquer, and so on
4. Pigments in an oil that polymerizes in the presence of oxygen and drying agents of alkyd, enamels, varnishes, and so on
5. Coatings formed by dipping in a hot melt of plastic such as polyethylene or acrylic
6. Coatings formed by using a powdered plastic and melting the powder to form a coating using many different TPs

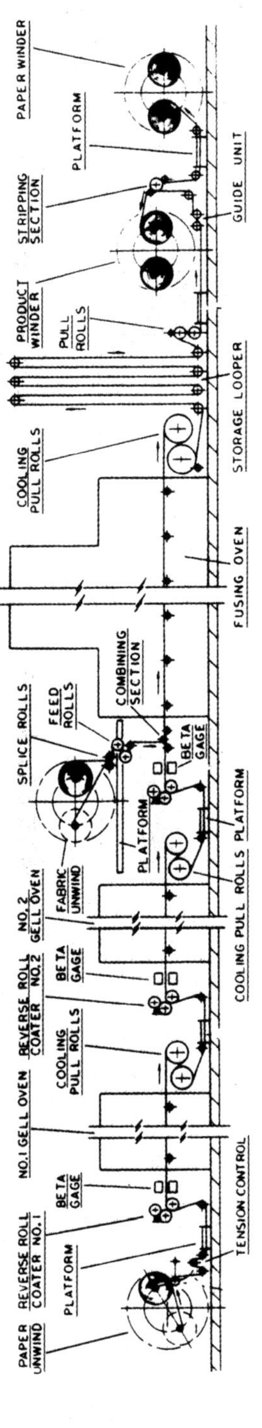

Figure 10.8 Cast coating line for coating by transfer from paper carrier.

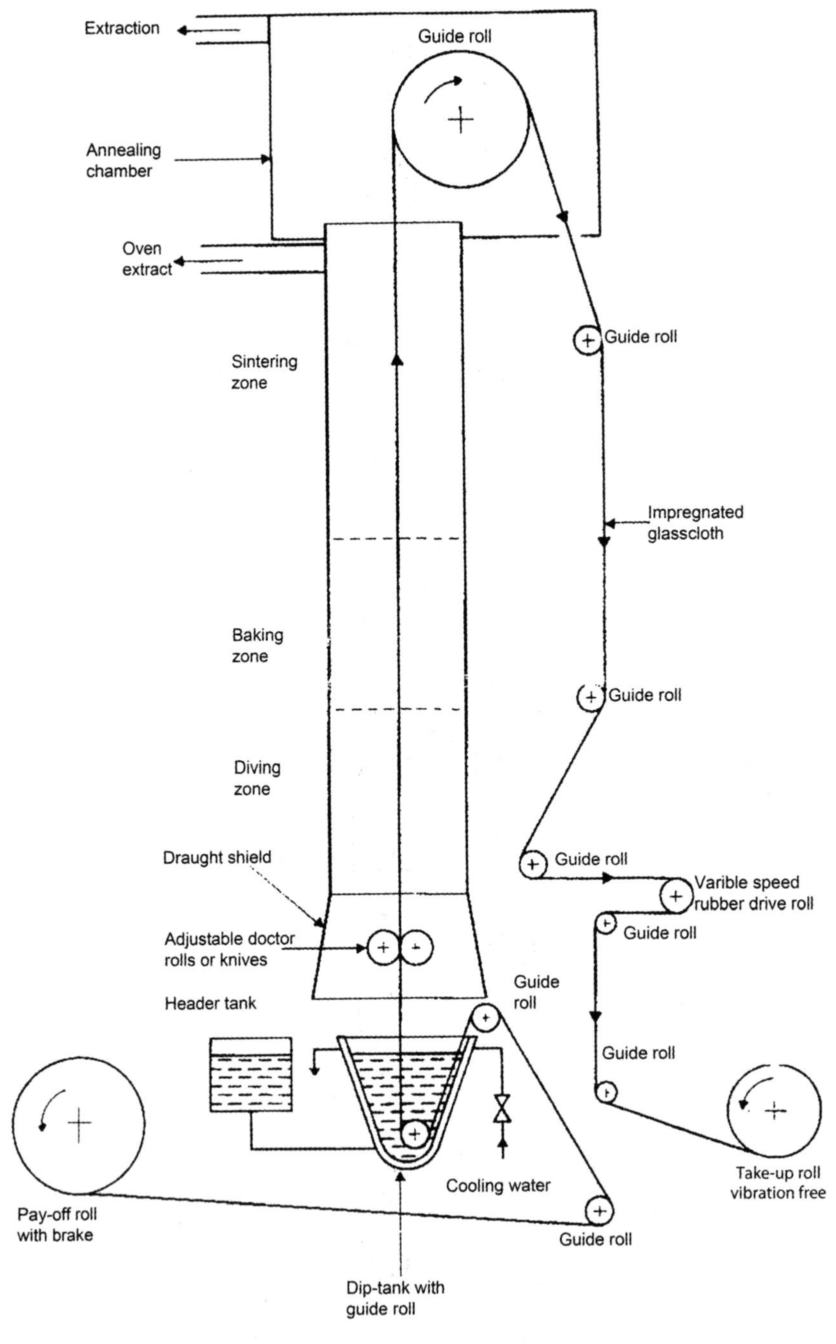

Figure 10.9 Fabric dip coating line.

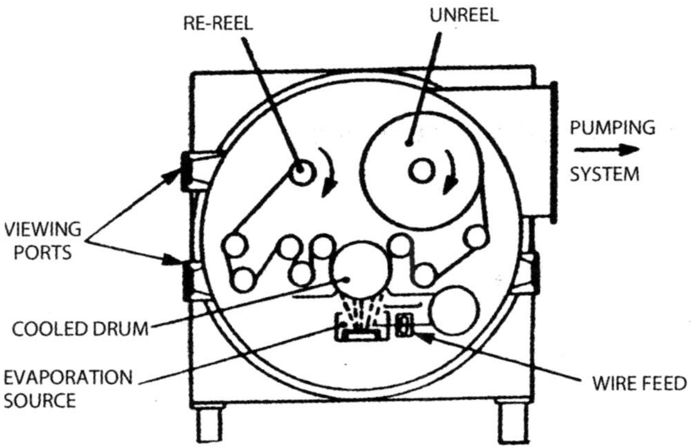

Figure 10.10 Example of a vacuum coater.

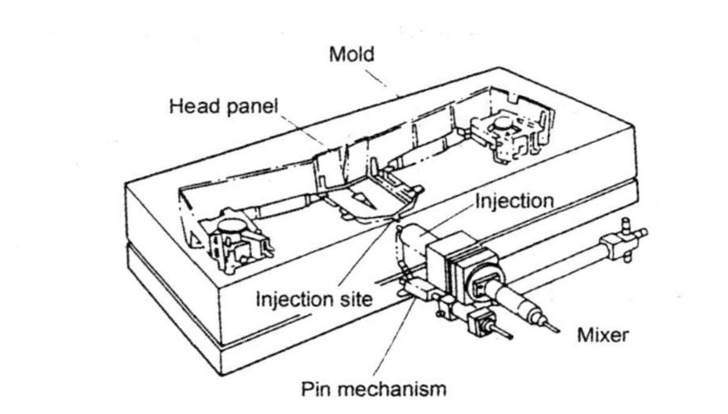

Figure 10.11 In-mold coating used in the reaction injection molding process.

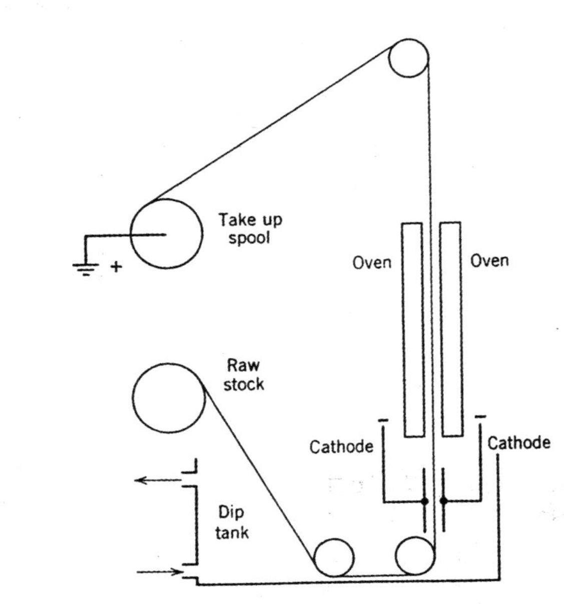

Figure 10.12 Electrodeposition for application of coating to magnet wire or strip.

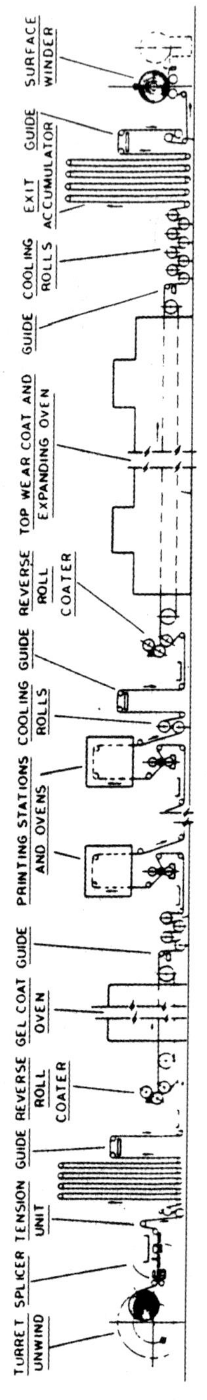

Figure 10.13 Floor covering coating line.

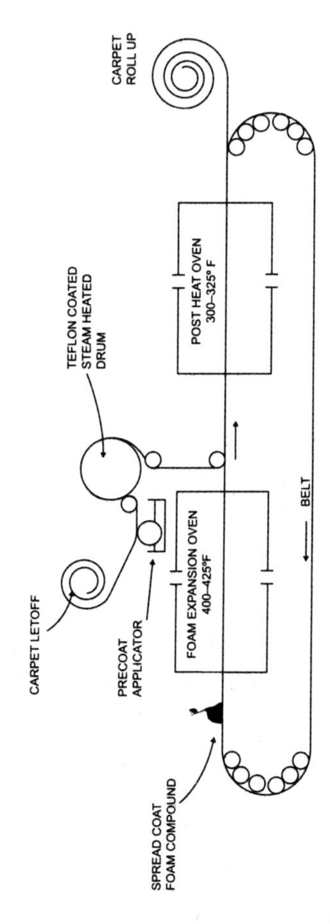

Figure 10.14 Foam plastic carpet backing coating line.

Spray method	*Transfer efficiency (%)*
Air atomization	35
Air atomization with electrostatic charge	60
Airless atomization	50
Airless atomization with electrostatic charge	70
Disk-and-bell atomization with electrostatic charge	90

Table 10.22 Examples of spray coating methods related to transfer efficiency

Equipment for coating lines can be associated with end-use markets, with some overlapping. Substrates or web-handling characteristics distinguish the differences among thin plastic film, paper, and paperboard combinations. Flexible packaging extrusion lines are using progressively thinner substrates of polyester, oriented polypropylene, metallized materials, and so on. Thin snack-food substrates require minimum tensions to assure that preprinted webs are not distorted.

Features include DC regenerative unwinds and in-feed holdbacks for precise and low-level tension; direct or reverse gravure for aqueous PVDC and other plastics; coating; infrared preheating; and vacuum rolls for web control. The concept of tandem operations or coating two sides of a substrate continues to expand to many flexible packaging lines that produce all kinds of combinations (different plastics, paper, aluminum foil, wood, steel sheet, etc.) and coating a plastic for heat-sealing. Higher operating-line tensions are used in producing structures with paper for granulated or powdered mixes and freezer or sugar-wrap materials. Different plastics, such as ionomers, acrylics, nylons, EVOHs, and EVAs, may be part of a converter's inventory of resins.

The combination of aluminum foil and barrier resins extends existing technologies to create lines with triple or quadruple (or more) coating systems and includes coextrusion in one or more locations.

FILM SOLIDIFICATION

When the coating is applied to the surface, the volatile liquid evaporates, leaving the nonvolatile binder-pigment combination as a residual film; it may or may not require a chemical conversion to an insoluble condition. Small amounts of additives are often included to improve application, pigment settling, drying, and film properties. Most binders are either high-molecular-weight, nonreactive plastics or low- to medium-molecular-weight, reactive plastics capable of being further polymerized via chain-extension or cross-linking reactions to high-molecular-weight films (chapter 1).

Most coatings are manufactured and applied as liquids; they are converted to solid films once they are on the substrate. Powder coatings are applied as a solid powder, converted to a liquid on the substrate, and then formed into a solid film.

Coating films are viscoelastic, so their mechanical properties depend on the temperature and the rate of stress application. Their behavior approaches the elastic mode with increased tensile strength to failure (breakage) or decreased elongation to failure, and with a more nearly constant modulus as a function of stress when the temperature decreases or when the rate of application of stress increases (chapter 19). The shifts can be especially large if results are compared above and below the glass transition temperature (T_g; chapter 1). Below T_g, the coatings have an elastic response and are therefore brittle; they break if the relatively low elongation to failure is exceeded. Above T_g, the viscous component of the deformation response is larger; the films are softer (lower modulus) and less likely to break during forming. Caution is required in considering the relationship of T_g to formability because some materials, such as acrylic and especially PC, are ductile at temperatures far below T_g. Above T_g, the modulus is primarily controlled by the density of the plastic cross-linkages.

After a coating is applied, solvent evaporation and rheological factors contribute to the solidification or curing of the coating film. Solvent initially evaporates from the surface of the film at about the same rate as it would in the absence of a binder. As the film solidifies, evaporation slows down because the diffusion rate to the surface is usually slower than the evaporation rate. Lacquer films do not cure by chemical reaction to achieve the required hardness and toughness. They just dry by solvent evaporation and depend on the high MW of these TP materials to provide the required performance. Latex paints behave in a similar manner.

COATING METHODS

Many different methods are used to apply plastic coatings to substrates of all sizes and types, ranging from the simple to the complex. They are generally composed of one or more plastics, a mixture of solvents (except with powder coatings), commonly one or more pigments, and frequently several additives. Coatings can be classified as TPs or TSs. Coating methods are categorized in different ways by the different industries that require them.

Traditional paints contain a vehicle, a solvent, and a pigment. Some are applied by spraying or dipping. Other systems involve heating parts and spraying them with a dry plastic powder that coalesces on the hot part to form a film. The differences among the various coating systems are the mechanism of film formation and the type of plastic being applied. Many important details are involved in surface preparation and in application techniques.

Both solvent-borne and aqueous paints are used. Paints are usually classified on the basis of the binder (vehicle) used. The most often used are (1) acrylics (aqueous acrylic emulsions, solvent-borne enamels, melamine, and other modified acrylic emulsions); (2) PURs (aqueous and solvent-borne); (3) alkyds and modifications; (4) epoxies and modifications; (5) polyesters;

(6) vinyls and modifications (latex or solvent-borne); (7) nitrocelluloses (solvent-borne); and (8) polyamides (solvent-borne).

The distinction between paints and enamels is not straightforward. However, enamels generally contain higher MW binders and are formulated with lower solids concentration. They are also formulated at lower pigment–binder ratios to create a superior gloss.

Lacquers differ from paints and enamels because they are compounded with TPs, which are soluble plastics of much higher MW and low chemical reactivity. Film formation occurs by solvent evaporation. Conventional lacquers are normally solvent-borne. Dispersions of plastics in water, latexes, or organic vinyl liquids (organosols) yield soluble films of TP; they also qualify as lacquers. Plastisols are dispersions of finely divided vinyl in plasticizers that are nonsolvent at room temperature but are good solvents at high temperatures. They are stable under normal storage conditions and can be coalesced into films at elevated temperatures.

Some plastic products that require painting may need special considerations because of their surface conditions. Some plastics may be sensitive to certain solvents, so take care to understand the situation.

Plastic coating substrates represent a big business. The substrates may be (1) films such as plastics and aluminum foils; (2) papers; (3) fabrics that are woven or nonwoven or both; (4) concrete, stone, and other types of masonry; (5) panels of wood, steel, and so on; (6) profile shapes made from different materials; (7) tanks and storage bins; and so on. The coating material provides many properties required to make the substrates more useful in commercial and industrial applications. Considerations in selecting the plastic coating include such factors as chemical environment, mechanical properties, processing characteristics, and costs.

Films are coated to extend the utility of the substrate by improving existing properties or adding new and unique properties. The coatings can provide heat sealability; impermeability to moisture, water, vapor, perfumes, and other gases; heat and ultraviolet barriers; modified optical or electrical properties; altered coefficients of friction; and a tendency toward blocking.

Coatings are different from laminations of two or more films. Laminates vary in construction: plastic film to aluminum foil, two or more plastic films combined, plastic film to paper to plastic film, paper to plastic film to paper, and so on. With plastic films, the coatings are usually thinner than the base film. Coatings are generally 0.05 to 0.2 mil (1.3 to 5.1 μm) thick. In laminations most films are at least 0.25 mil (6.4 μm) thick, and more commonly 0.5 to 2 mil (13 to 51 μm) thick.

Different desirable properties of a fabric can be supplemented by plastic coating. The fabrics provide at least tensile and shear strengths with elongation control. Coatings can protect the fabric, reduce porosity, provide decorative effects, and other benefits. Coated fabrics are designed for specific applications. The three major considerations are the physical environment, the chemical environment (water, acid, solvents, and so on), and cost. Impregnation is the process of thoroughly soaking and filling the voids and interstices of the substrate (as well as wood and paper) with the plastic coating. The porous materials generally serve as reinforcements for the plastic after the coating treatment.

Processing is dictated by the properties of the substrate and the coating. The viscosity of the coating must permit flow around the yarn or fiber surface. In extrusion and calendering, pressure and heat fluidize the coating. In other processes, solution or dispersion can reduce viscosity.

Wall coverings, upholstery, and apparel are examples of decorative coated fabrics. Inks are applied with one or more gravure printers to correct the color or to add a pattern. Relief patterns are obtained by applying heat and pressure with embossing rolls.

There are leather substitutes that are designed to imitate the appearance of leather with its surface grain. This is accomplished by coating substances that are capable of forming a uniform film. Plasticized PVC first met this requirement during the 1940s. When plasticized PVC (solid or foam) is coated onto a substrate, it produces a leather-like material called vinyl coated fabric. It exhibits high density, very low water-vapor permeability, cold touch, poor flex endurance, and poor plasticizer migration. But it has good scratch resistance and colorability as well as being inexpensive.

PUR coated fabrics, developed in the 1960s, were an improvement. PUR is coated on woven or knitted fabrics. With a T_g below 32°F (0°C), PUR is very flexible at room temperature without a plasticizer. Another important characteristic is that its molecular structure allows water-vapor permeability. In addition, the solvents normally used for a PUR will permit coagulation by a nonsolvent with formation of a porous structure. The result is increased flexibility and water-vapor permeability.

Drying a cast PUR solution to form a film that is laminated onto the substrate will produce ordinary PUR-coated fabrics. Significant improvements in appearance, feel, and grain are accomplished by using a brushed fabric as the substrate. It is laminated with a cast PUR film. Alternatively, an organic solvent solution of PUR is applied to a brushed, woven fabric immersed in a nonsolvent bath for coagulation.

The poromerics are also called synthetic leather. They were developed during the 1960s as an improvement over fabrics coated with leather-like coatings, whose applications were limited by the properties of the knitted or woven substrate. Poromerics use a nonwoven fabric impregnated with plastic, which thereby creates a substrate resembling leather. Fine fiber construction provides the desired softness. Prepared with PUR, the poromeric coating layer corresponds to the grain of the leather.

Historically, smoke and the resulting toxic fumes caused by the burning of a flammable substrate were part of any fire, regardless of whether a fire-retardant treatment was applied. What was needed was to smother the fire and thus stop the generation of toxic smoke and prevent further damage to the substrate. Intumescence coatings were developed over a half-century ago by the US Navy for use on ships. Industry projects developed different types of water-resistant intumescent coatings. These intumescent coatings, when subjected to fire, form a char between the substrate and the fire source. The basic product becomes flameproof.

Intumescence coatings provide the most effective fire-resistant system available, but originally they were deficient in paint color properties. Since, historically, the intumescence-producing chemicals were quite soluble in water, coatings based on those chemicals did not meet the shipping-can stability, ease of application, environmental resistance, or aesthetic appeal required of a good protective coating.

COATING EQUIPMENT

There are different methods used; examples are shown in Figure. 2.4 and Figure 2.5. Each has its performance advantages and cost benefits. Coating equipment is used to apply a surface coating, a laminating adhesive, and any compounds for saturation or impregnation (or both) of a fabric. The equipment has three basic components: the coating head, a dryer or other coating solidification unit, and web-handling hardware (drives, winders, edge guides, controls, etc.). It can generally coat various substrates in roll or sheet form.

Coatings can be applied directly to the substrate or transferred to the substrate from another surface, such as a roll. Transfer from another surface is used when the substrate is sensitive to the coating material, when it may be damaged by exposure to oven temperatures, for special secondary operations such as applying pressure-sensitive labels, and so on.

During its application, the coating must be sufficiently fluid to be spread into a uniformly thin layer across a web. Coatings can be applied as solutions in organic solvents, as aqueous solutions or emulsions, or as molten or softened solids. Solutions and emulsions require drying to obtain solid coatings. Cooling solidifies hot melts. Some coatings may be applied as reactive liquids and then polymerized by infrared or heat.

Heat and mass transfer take place simultaneously during the drying process. The heat is transferred by convection in air dryers, by radiation in infrared dryers, and by conduction in contact-drum dryers. The drying equipment usually has a means to remove and recirculate the vapor with heat-exchange equipment to conserve energy.

The coating head accomplishes two functions. It applies the coating to the substrate, distributing it uniformly in metered amounts over the surface. Most coaters fall into the following categories: roll, knife, blade, or bar. There are also extrusion or slot-orifice coaters.

Roll coaters, the most widely used kind of coater, are subdivided by their construction, such as direct, reverse, gravure, or calender. Examples of coating equipment include the following: roll coaters, knife bar coaters, curtain coaters, and equipment for coil coating, vacuum coating, spray coating, floc coating, electrodeposition coating, powder coating, fluidized bed coating, electrostatic coating, electrostatic fluidized coating, flood coating, microencapsulation coating, pinhole-free thin coating, and radiation curing (1).

ROLL-COAT FINISH

Referred to as "roll-coat finishes" because they are applied to coiled metal by the reserve roller-coating technique (similar to offset printing), these finishes have grown into a sophisticated group of materials since their inception about 80 years ago and are now offered in a wide variety. Their primary advantage is that they can withstand metalworking operations without any resulting surface damage. Thus they can be applied before product fabrication, which eliminates finishing steps afterward and can thereby cut costs.

With the wide range of resins, there are types of roll-coat finishes that are extremely flexible, capable of taking very severe forming operations with no cracking or loss of adhesion. Consequently, they are being used for applications involving rigorous bends, which before prohibited the use of precoated metal for lack of finishes with enough formability. One such material is a vinyl coating. It can satisfactorily withstand one of the most troublesome bends, the zero radius or back-to-back bend. There are also flexible acrylics and polyesters.

Another advantage offered by these materials is the broad range of decorative effects they can achieve, which also has been boosted by the wider variety of resins available.

Because roller coating is a high-speed operation, these roll-coat finishes have to cure quickly. Therefore, modifications have been made to upgrade the performance of plastic resins in this area. An average baking cycle today is 60 seconds at 500°F (260°C), in which time over 200 ft of coil is coated. With coil coaters operating at even greater speeds, the resins have to cure in shorter bakes. Other properties of the roll-coat finishes that have been continually improved a great deal are their exterior durability, chemical resistance, and color retention.

SPREAD COATING

In spread coating, the material to be coated passes over a roller and under a long blade or knife. The plastic coating compound is placed on the material just in front of the knife and is spread out over the material. The thickness of the coating is basically regulated by the speed at which the material is drawn under the knife and the position of the knife. In roller coating, two horizontal rollers are used. One roller picks up the plastic coating solution on its surface and deposits it on the second roller that, in turn, deposits the coating solution on the supporting material. The usual coating material is a plastic melt but plastics in the form of fine powders are also used.

FLOATING KNIFE COATER

This coating machine applies a uniformly controlled amount of forming, sizing, or other desirable material to a web or a sheet of substrate. The choice of coater (spread, spray, roll, dip, and air knife) depends on the type of coating and the substrate. Other factors such as solvent removal, drying, and production rate must be considered.

Spread coaters include the knife or bar coaters that scrape off a heavy layer of coating liquid to the desired thickness. The floating blade coater depends on web tension and blade contour to control thickness, whereas the knife-over-roll configuration (Fig. 10.5) allows you to set the knife at a fixed distance from the roll. Modifications of knife contour control coatings of various viscosities and rheologies exist.

A unique form of spread coater operates by applying an excess of coating and then metering with a transverse rod helically wrapped with a wire (or rod). The gauge of the wire governs the thickness of the remaining coating (at constant solids content). A coarse wire gives heavier coatings, while a fine one leaves thinner films. The rod is most often used for thixotropic solutions and dispersions; dilatant liquids do not perform well with this method.

Application of fluid coatings to a web by spraying is usually accomplished with multiple spray heads mounted on an oscillating carrier. Mount the spray heads so the patterns overlap, and move them across the web to lay down a uniform coating.

There are many types of roll coaters available; perhaps the most successful is some version of the reverse-roll arrangement. The reverse roll is so called because the roll rotates counter to the substrate travel. This allows you to control coating thickness by adjusting the gap between either the metering roll or applicator roll or both. The reverse roll coater works best at applying coatings that are thixotropic or at least Newtonian (chapter 1). Coatings of a dilatant nature generally run at lower speeds because of the high shear between the applicator roll and the substrate.

In some instances, where both sides of a substrate must be coated, it is best to dip the substrate directly into the coating and remove the excess to leave the desired thickness. This is achieved by passing the coated substrate between two rolls or two wire-wound rods. Fusing two-sided coatings of sticky substances is generally difficult, and production speeds are usually very low.

Air knife coating machines are used for applying water dispersions or an emulsion, where solvent loss and resultant surface skinning is not a problem. By using an arrangement similar to the one depicted in Figure 13.39, you can apply a smooth, uniform coating.

FLUIDIZED BED COATING

In fluidized bed coating, the object to be coated is heated and then immersed in a dense-phase air fluidized bed of powdered plastic; the plastic adheres to the heated object and subsequent heating provides a smooth, pinhole-free coating.

SPRAY COATING

Spray coating is used before and after assembly, the latter particularly if the product is already assembled and has a complex shaped and curved surfaces. Many different types of spray equipment are in use to handle the different forms of paints. They are classified by their method of atomization (airless, air, rotary, electrostatic, etc.) and by their deposition technique (electrostatic or nonelectrostatic, flame spray, etc.). Spraying techniques may fall into several of these categories. They range from simple systems with one manual applicator to highly complex, computer-controlled, automatic systems. They can incorporate hundreds of spray units. Automatic systems may have their applicators mounted on fixed stands, on reciprocating or rotating machines, on robots, and so on.

FLAME SPRAY COATING

Flame spray coating consists of blowing a powder through a flame that partially melts the powder and fuses it as it contacts the substrate. The part's surface is preheated with the flame, usually to about 400°F (204°C) when using polyethylene. The usual approach is to coat only a few square meters at a time so the temperature can be controlled. The flame is then adjusted. When coating is completed, the powder is shut off and the coating is postheated with the flame. Flame spraying is particularly useful for coating products with surface areas too large for heating in an oven. Disadvantages are the problems associated with an open flame and the need for skilled operators to apply the coating.

Powder Coating

Powder coating is a solventless system; it does not depend on a sacrificial medium such as a solvent, but is based on the performance constituents of solid TP or TS materials. It can be a homogeneous blend of the plastic with fillers and additives in the form of dry, fine particles of a compound similar to flour.

Advantages of powder coating include minimum air pollution and water contamination, increased performance with coating, and consequent cost savings. It has many of the same problems as solution painting. If not properly formulated, the coating may sag (particularly if it is thick), show poor performance when not completely cured, show imperfections such as craters and pinholes, and have poor hiding with low film thickness. Various methods are used to apply powder coatings.

Electrostatic Spraying

Electrostatic spraying is based on the fact that most plastic powders are insulators with relatively high volume resistivity values. Therefore, they accept a charge (positive or negative polarity) and are attracted to a grounded or oppositely charged object (which is the one being coated).

Coil Coating (Metal Coating)

Coil coating processes involve high speed (at least 500 ft/min) and continuous mechanized procedures for paint coating one or both sides of a coil of sheet metal. Coating equipment, metal cleaning, and new paint formulations provide ease of formability with environmental durability. The basic operations in the process involve unwinding steel coil, chemically pretreating steel, reverse roll-coating paint, baking paint, applying additional coatings in certain processes, cooling coated metal, inspection, and rewind coil.

The National Coil Coating Association in Cleveland, organized in 1962, has already been very active in such operations as developing industry standards, exchanging technical information, preparing technical manuals, and keeping records of sales growth.

The first extensive market for this product was for venetian blinds, followed by metal awnings, metal sidings, automobile trims, light reflectors, luggage, metal doors, and other similar products.

PROPERTY

Plastic coating materials have been exposed to all kinds of performance tests and environments to meet the many different requirements that exist in the many different applications. Figures 10.15 to 10.20 show a few properties of coatings when in severe environments. What follows is information that highlights some of the properties and tests that influence the performance of coatings starting with Table 10.23 (chapter 22).

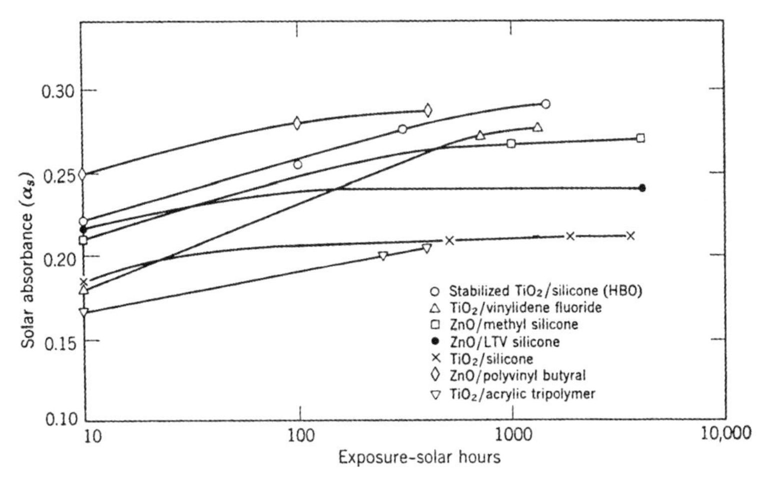

Figure 10.15 Vacuum-ultraviolet radiation effects on organic coatings.

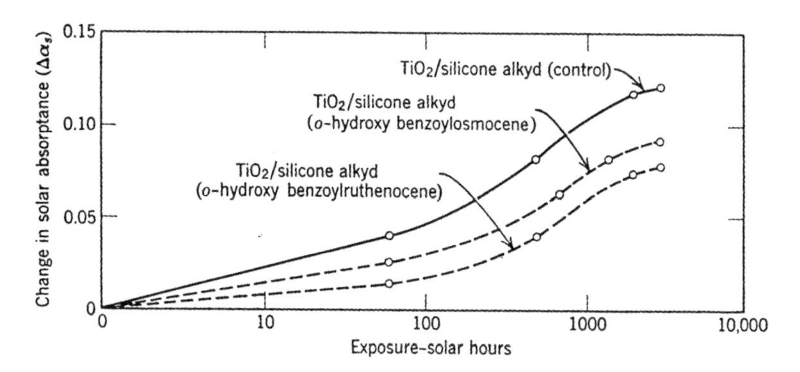

Figure 10.16 Vacuum-ultraviolet radiation effects on stabilized organic coatings.

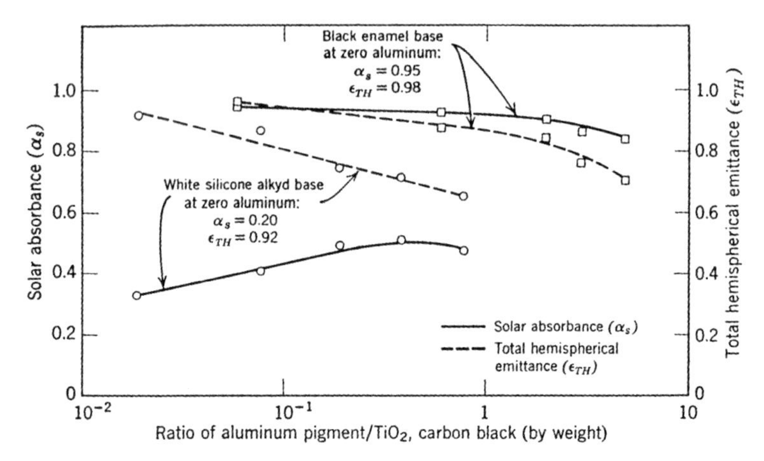

Figure 10.17 Relationship between solar absorbance, total hemispherical emittance, and pigment ratios.

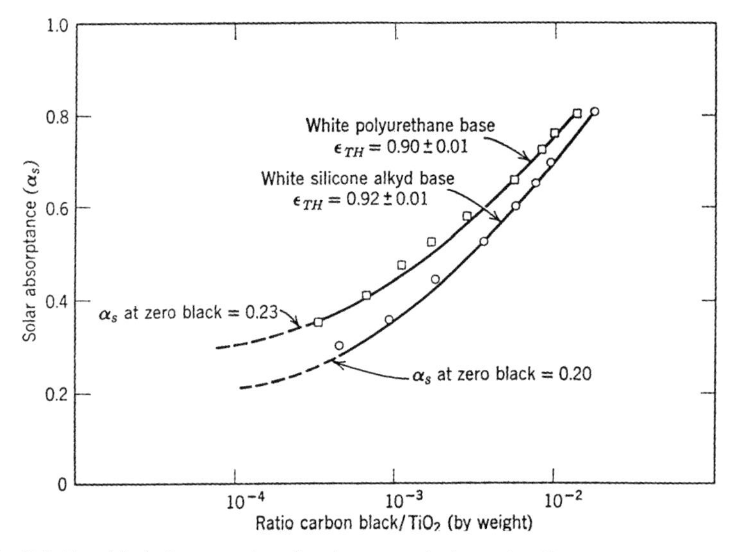

Figure 10.18 Relationship between solar absorbance and pigment ratios.

CORROSION AND CHEMICAL RESISTANCE

Corrosion- and chemical-resistant paints are used to protect industrial products and structures from attack by the weather and corrosive materials. A wide range of protective coatings has been developed to meet various different service conditions, methods of application, drying schedules, and other performance and economic requirements. Some of these coatings are used without

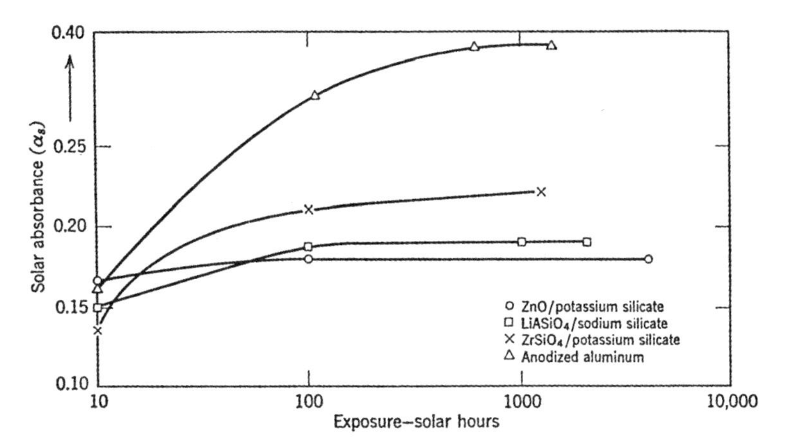

Figure 10.19 Vacuum-ultraviolet radiation effects on inorganic coatings.

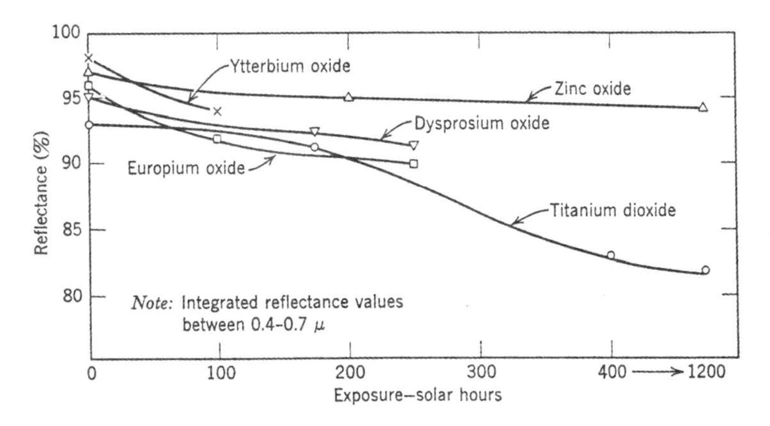

Figure 10.20 Effects of vacuum-ultraviolet radiation on pigments.

pigment and are based on synthetic resins and rubbers. Others contain special corrosion-inhibiting pigments; their binders are materials such as drying oils, varnishes, alkyds, and epoxies.

Another group of these coatings is based on bituminous materials. Although they are designed to provide a protective barrier between the surface to be protected and the corroding elements or materials, they utilize two different methods of protection. For example, clear coatings and those based on bituminous materials protect essentially by providing a barrier that is of sufficient thickness or resistance (or both) to service conditions to prevent the corroding environment from reaching the protected surface.

However, corrosion-resistant coatings based on drying oils, varnishes, and other products are quite permeable to water and oxygen. These coatings depend almost entirely on their inhibiting

	Amine-Alkyd	Polyester (Oil-Free)	Vinyl-Alkyd	Solution Vinyl (PVC)	Organosol	Plastisol	Straight Epoxy	Epoxy-Ester	Thermoset Acrylic	Silicone Alkyd	Silicone Acrylic	Silicone Polyester	Polyvinyl Fluoride	Polyvinylidene Fluoride	Polyvinyl Fluoride Film Laminate	Polyvinyl Chloride Laminate	Acrylic Film Laminate
Hardness	2	1	2	2	2	3	1	2	2	2	2	2	2	2	2	2	1
Adhesion	2	1	2	1	1	2	1	1	1	2	2	2	2	2	2	1	1
Flexibility	3	2	2	1	1	1	2	2	2	3	3	2	1	1	1	1	1
Mar resistance	2	2	2	2	3	3	1	2	1	2	2	2	3	3		2	2
Gloss (85 units plus 60°)	1	1	2	2	3	4	1	2	1	1	1	1	4	4	4	4	2
Fabricability after aging	4	1	2	1	1	1	3	2	2	3	3	2	1	1	1	1	1
Humidity resistance	2	1	1	1	1	1	1	1	1	2	2	1	1	1	1	1	1
Grease and oil resistance	2	2	2	1	1	1	1	2	2	2	2	2	1	1	1	1	1
General chemical resistance	3	2	3	2	1	1	1	2	2	2	2	2	1	1	1	1	1
General corrosion resistance (industrial atmospheres)	2	2	2	2	1	1	1	2	2	2	2	2	1	1	1	1	1
Exterior durability (pigmented)	2	2	3	2	2	2	4	4	2	1	1	1	1	1	1	2	1
Exterior durability (clear films)	3	3	3	3	3	3	4	4	2	2	2	2	1	1	1	4	1

Ratings: 1 = excellent, 2 = good, 3 = fair, 4 = poor.

Table 10.23 Plastic coating property guide

pigment to control corrosion. In addition, the electrolytic resistance of coatings is an important factor in their ability to inhibit corrosion.

Many different factors must be considered in the development of a suitable coating system. In many cases, combinations of corrosion-inhibiting primers obtain the best results and produce highly resistant finished coats. However, in all cases it has been demonstrated quite clearly that the paint system must have sufficient thickness to obtain adequate protection. It is generally believed that the minimum thickness of the paint system should be 5 mils.

FIRE RETARDANT

The majorities of paint binders are organic in nature and hence are inflammable. Incorporating additives such as chlorine, bromine, nitrogen (bound in particular ways), phosphorus, or silica can reduce the inflammability of organic material. Coatings based on noninflammable binders include lime washers, cement paints, and silica paints.

Inorganic pigments are generally noninflammable. Antimony trioxides and zinc borates are particularly effective. Combinations of organic and these inorganic pigments provide fire-retardant characteristics.

Very few clear fire-retardant coatings are available. They are generally two-part systems using epoxy or urethane resin. There is a transparent PUR type for use on wood and metal parts such as walls, ceilings, building fixtures, and furniture. When subjected to fire, it swells, creating an insulating surface of charred foam that prevents further burning. This is fundamentally a practical application for what was developed during the Korean War as fire walls for use in different military vehicles.

These coating are not sensitive to moisture, as were the previous types. A dry coating can expand 166 times, or go from 6 mils thickness to 1 in. This system provides a major step forward in saving lives and property from fire.

In the meantime, this industry continues to target the development of a fire-retardant coating that does not cause the usual loss in other properties and does not significantly increase cost. Of course the real push for these coatings will occur if state and local governments adopt more stringent building codes.

INTUMESCENT COATING

Intumescent coatings, or coatings that bubble and foam to form a thermal insulation, have been used for many years. Such coatings cannot otherwise be differentiated from conventional products. Thereupon, however, they decompose to form a thick, nonflammable, multicellular, and insulative barrier over the surface on which they are applied. This insulative foam is a very effective insulation that maintains the temperature of a flammable or heat-distortable substrate below its ignition or distortion point. It also restricts the flow of air (oxygen) from the substrate fuel.

It is estimated that millions of gallons of alkyd and latex-emulsion intumescent paints are used annually. These coatings provide the most effective fire-resistant system available, but originally they were deficient in paint properties. Since, historically, the intumescence-producing chemicals were quite soluble in water, coatings based on those chemicals did not meet the can stability, ease of application, environmental resistance, or aesthetic appeal required of a good protective coating.

In time to meet market requirements, Monsanto Company developed a new water-resistant, phosphorus-based intumescence catalyst. This new commercially available product, Phos-Chek P/30, can be incorporated (with other water-insoluble reagents) into water-resistant intumescent coatings of either the alkyd or latex emulsion type. These intumescent coatings, formulated according to the manufacturer's recommendations, are described as equivalent to conventional products in coating properties. They also provide permanent fire resistance to the substrate on which they are applied.

HEAT RESISTANT

Different types of coatings are available to meet the different heat resistant environments. There are coatings that can be exposed to high temperature intermittently. The film is subjected to mechanical forces by differential expansion and contraction. Limiting factors involve breaking the bond to the substrate and the effect at elevated temperatures of oxygen attack on the coatings (when in an oxygen-rich atmosphere).

Inorganic polymers and semiorganic polymers are the binders that show the best heat resistance. Extensive research and development have been undertaken to study polymers containing boron, nitrogen, phosphorus, silicon, and so on. The esters of silicon, titanium and the silicones are examples of binders that are suitable for continuous use at temperatures above 150°F (66°C). For the best practical results, they are pigmented with leafing aluminum. This composite protects the organic part of the binder from oxygen attack and also forms a metal-ceramic complex.

For temperatures up to 150°F, many binders may be used to obtain a useful life. Short oil phenolic varnishes, oxidized rubber, and melamine/alkyd resin stoving finishes (where the alkyd is a saturated fatty acid/isophthalic acid type) have been used in various coating systems.

THERMAL CONTROL

Since 1960, passive thermal control of space vehicles and their components has emerged as an area of increasing importance among the space sciences. This area is destined to achieve greater stature as our ventures into space become longer in duration and complexity. Indicative of this importance is the research now being devoted to space exploration.

In contrast to active thermal control, passive thermal control offered the advantages of no moving parts, hence the absence of mechanical failure, and a considerable weight savings. The fundamental parameters in controlling the space vehicle's temperature by passive means are the optical characteristics of the surface of the spacecraft, that is, the solar absorptance and the emittance. However, in order to function as a thermal control surface, a coating must be optically stable in the space environment, especially with respect to ultraviolet radiation, particulate radiation, high vacuum, and temperature. In addition, the properties of the coating, such as flexibility, adhesion, ease of application and ability to air-cure could not be overlooked.

The major effort over the past several years has thus been directed toward the preparation of organic and inorganic coatings with desirable temperature-control properties. At the same time, researchers pursued a critical evaluation of the factors of the space environment and performed laboratory research involving the effects of these factors on the optical and physical properties on coatings and coating materials.

In an attempt to simulate the environment of outer space, the research worker was faced with many unknown factors. The definition of the space environment has been and is presently a major area of scientific research in itself. This lack of overall knowledge, in addition to the economics involved in simulating the space environment in its entirety, translated into compromises in the exposure chambers.

Although white organic coatings have been shown to degrade, primarily by ultraviolet radiation, their flexibility, adhesion, ease of application, air-drying characteristics, and high emittance have persuaded researchers to spend considerable time and effort studying them. Inorganic coatings, however, offered the promise of stability to ultraviolet radiation and high vacuum, but presented problems in adhesion and flexibility and involved heat cures. The effects of particulate radiation, although not forgotten, remained somewhat in the background.

Organic formulations have been developed with air cure properties and optical properties. Figure 10.15 shows typical coatings. Pigment research, with regards to stability, compatibility, index of refraction, particle size, purity, and other properties, has shown ZnO, TiO, and ZnS to offer the most promise. The most stable vehicles in combination with these pigments involved the silicones and their variations, acrylics, silicone alkyds, and certain fluorocarbon polymers.

With the synthesis of colorless ultraviolet radiation absorbers, such as the derivatives of osmocene and ruthenocene, it has been possible to enhance the stability of otherwise borderline acceptable thermal control coatings to the point where these coatings become candidate materials (Fig. 10.16). The derivatives of ferrocene, although enhancing the stability of certain white coatings, gave a high initial performance due to its inherent red coloration. The derivatives of osmocene and ruthenocene, however, were white to light yellow.

The controlled addition of pigment to both basic white and black formulations has provided design engineers with a series of coatings with a wide range of performances for specific applications. These formulations are shown in Figures 10.17 and 10.18.

ELECTRICAL INSULATING

A new and revolutionary idea has been explored for the application of enamels to copper (or aluminum, etc.) magnet wire (Fig. 10.12). It is electrodeposition, a process of coating copper wire in an aqueous solution by an electric current. In this method, an organic resin particle is dispersed in water in such a manner that it has an electrical charge. Then, under the influence of an electrical field, it is attracted to a positive electrode, the copper-wire anode, and deposited onto it.

This method has been used for years in the electroplating of metals. It was not until the resin chemist was able to formulate superior water-soluble, or water-dispersible, resins, that these basic principles could be applied to the coating industry.

The process of electrodeposition produces an unusual phenomenon. The resin plays a dual role in the process. The material is a conductor at one point, and a few minutes later it is an insulator. The objective of the program was to resolve these two paradoxical processes. Perhaps it might appear that with this approach, in which the insulator is water soluble during its processing, the coating would still have an affinity for water. However, the coated enamel has lost its water solubility through the baking cycle during which it has become an insoluble, TS resin. Correct formulating principles permit the resin to perform this dual role.

The principal advantages of electrodeposition over conventional methods are as follows:

1. Faster production rates. Production of coated wire can be increased four to six times per hour. Only one passage of the wire through the dip tank is required. For example, a processing wire mill will have to run #18 AWG copper magnet wire through the dip tank only one time—rather than four to six times—to obtain heavy build.
2. Superior properties through formulation of higher MW materials. The resin chemist will be able to develop unusually long chain polymers because water is used in the system.
3. Elimination of offensive odors, especially cresylic acid fumes that are common in cities.
4. Complete and uniform coverage. In electrodeposition, the charged negative particles are attracted to positive metal surfaces. This attraction will continue until the entire metal surface is covered.

The phenomena of electrodeposition, as applied to a protective coating application method, are complex. Although the method employs theories established many years ago, the practical application of these principles has introduced variables that complicate the deposition. To accomplish deposition by this method, it is necessary to have a water-soluble resin (or water-dispersible resin or emulsion) and a dip tank with two electrodes. When the current is turned on, the following reactions occur: electroysis, electrophoresis, electrocoagulation, and electroosmosis. All four of these reactions usually occur simultaneously during the deposition of a resin. They play an important role in the correct application of the resin to the wire.

SIMULATED SERVICE TESTS

Many different service tests have been used throughout the centuries, and more tests will be developed in the future. This way of life is directly related to progress. Of course the real test is when parts are in use. Many of the laboratory tests described throughout this chapter and book will provide useful ideas for conducting simulated service tests. This section presents a review on how to evaluate coated fabrics. Different tests can be used to evaluate elastomer-coated fabrics. The following typical tests provide brief and limited examples of information on testing procedures.

ABRASION RESISTANCE

Abrasion is of primary importance with materials that are exposed to much abrasion, such as truck covers. Tests should be conducted after exposure to the liquid in which it will operate. Test values have only relative merit; projection of results to service can be difficult without experience, since many factors affect performance. Examples of tests follow.

In the Taber Abrasion Test, abrasion wheels and weights are used in various combinations to abrade one or both sides of a fabric. Values are reported in numbers of cycles needed to achieve either first exposure of the base fabric or to cause initial tearing. In another method, the abrader is run for a number of given cycles, and then the coating's weight loss is measured. A less accurate method involves judging the appearance of fabric after a given number of cycles.

In the DuPont Scrub Test, a specimen is vigorously scrubbed against itself between two moving jaws. The number of cycles needed to first expose the base fabric is reported. Values are usually determined for each side and for warp and filling direction. Visual inspection after a given number of cycles is not as accurate.

In the Wysenback Test, also called the oscillatory cylinder test, the fabric is subjected to unidirectional rubbing action under known conditions of pressure, tension, and abrasive action. To test specimens, they are placed in contact with an oscillating cylinder covered with an abrasive material. Values are reported for loss of breaking strength after 250 cycles of continuous abrasion under a 2 lb load (ASTM D1175).

ADHESION

Adhesion is particularly important in applications where a fabric is exposed to a great amount of flexing, pressure, and stretching. For fabrics with several plies, adhesion between plies should be tested. It is a particularly critical property in such products as life rafts, life vests, and other bonded inflatables. A testing example is the Scott test method, in which adhesion is tested by peeling coating from the base fabric, or by peeling two plies apart. A 2 in wide sample is inserted in jaws that separate at a given speed, most frequently 2 ipm or 12 ipm. Results are expressed in lb/in of width required to peel coating and fabric or two plies apart (ASTM D751).

ACCELERATED AGING IN HOT AIR

This is a useful measure of service life at atmospheric temperatures, and is particularly important for fabrics that must perform at elevated temperatures, such as automotive and heating equipment. The following paragraphs describe examples of the tests used.

HOT-AIR AGING

Samples are placed in a thermostatically controlled air-circulating oven and aged for specified temperatures and times. Fabrics that are not expected to perform in elevated temperatures are usually aged for 72 hours at 160°F (71°C). Fabrics used in automotive applications are aged for 72 hours at 200°F to 300°F (93°C to 149°C). If fabrics will be exposed to oils or other fluids during actual service, samples are immersed in suitable test fluids.

AGING IN STOP-AND-GO APPLICATIONS

Aging tests are interrupted and fabrics are allowed to dry between hot air tests. After aging, properties are compared with those taken before aging (ASTM D573).

ACCELERATED AGING IN OXYGEN

This testing establishes the behavior of fabrics that are exposed to air and sunshine in critical applications, such as life rafts, tarpaulins, and radome covers. In the oxygen bomb test, samples are placed in a bomb for a specified period of time at a temperature of 160°F and 300 psi (2 MPa) of oxygen pressure. After aging, most important physical properties are tested and checked against values obtained before aging (ASTM D52).

BLOCKING

This testing shows the tendency of certain elastomers, including natural rubber, to adhere to themselves. In a blocking test, a sample is folded and put under a 4 lb weight in an oven at 180°F (82°C) for 39 min. After cooling, the sample is unfolded and examined for adherence or peeling of the coating (ASTM D 1893).

BURST STRENGTH

This is an important guide in applications in which a fabric is exposed to high and sudden pressures, such as in gas regulators and controls. In the Mullen test, a sample is clamped across the orifice of a Mullen tester with a rubber diaphragm. Power is applied, and the pressure at which rupture occurs is recorded in gross and net psi values (ASTM D 751).

COMPRESSION SET

This measure of the permanent deformation of a fabric exposed to constant pressure or deflection can be important, particularly with coated fabric diaphragm assemblies in units with bolts or crimps. There are the test methods in which a sample is held under either constant load (method A) or constant deflection (method B) for a specified length of time in a specified constant temperature. The residual decrease in thickness, either a percentage of original thickness (method A) or a percentage of deflection under load (method B), is reported as the compression set. An interval of at least one week between removing the load and measuring the residual decrease in thickness or deflection is necessary to evaluate the permanent set (ASTM D 395).

CURL

Curl is an inconvenience in assembly, especially when using automatic assembly equipment. Fabrics that normally have the same coating on both sides should not curl; however, materials coated on one side almost always curl. There is a test method where a disc with a diameter of 2¼ in is allowed to rest for 24 hours at 70°F (21°C) with the side that shows the tendency to curl facing up. The distance from tabletop to the highest point of curl is reported as curl value. A curl of ¼ in or more is usually considered excessive for a diaphragm assembly.

ELONGATION

Elongation without rupture is desirable in diaphragms that pulse with a long stroke. In other devices in which dimensional stability is of utmost importance, such as metering devices, elongation of a fabric is undesirable. Test method samples are stretched in a Scott tester until they rupture. The amount of stretching that occurs between benchmarks at the time of rupture is the ultimate elongation of the fabric. Values are expressed as percentages of the original distance between the marks (ASTM D 751).

FLAME RESISTANCE, VERTICAL

A guide to flame resistance is important with any coated fabric used where a fire hazard exists. Several flame tests (UL) are available of which the severest and most widely used is vertical flame test. In this test, a sample measuring 2¾ in × 12 in is clamped between metal flanges and suspended in a metal cabinet to shield it from drafts. A flame of a standard size and a standard intensity is placed under the sample for 12 seconds. The burner is then removed without opening the cabinet. Three values are usually measured: (1) the length of time the flame continues to burn after the removal of the burner, (2) the length of time the glow continues, and (3) the char length, expressed as distance to which a charred sample will tear when subjected to a specified load.

FLEXIBILITY

This property is important in many lightweight fabric diaphragms used in delicate sensing devices. A testing machine takes one end of 1 × 3 in sample and places it in a clamp that will rotate the sample up to 90°. The other end of the sample actuates a balanced arm that gives readings up to 100 units. The stiffest samples give the highest readings; weights can be added to balance very stiff samples. Values (expressed in weights plus units) can be read at various degrees of rotation of the sample. Most commonly, the value at 30° rotation is read.

There is also a test in which a strip measuring 1¾ in wide is placed between two rollers with 1 in diameters held together by spring pressure. One of the rollers is slowly rotated. The overhanging sample is adjusted until it falls over in both directions of rotation when the framework is rotated clockwise and counterclockwise. The length of the overhanging portion, in mm, is the flexibility value.

HARDNESS

A measure of hardness is important in such products as rubber offset blankets. The hardness of a coating is usually measured before the coating material is applied to the base fabric. The popular Shore durometer is one method of testing. It usually gives valid readings only with unsupported elastomers of at least ¼ in thickness. On coated fabrics it does not produce reliable, absolute values because instrument registers some of the hardness of the base fabric.

A Shore durometer measures resistance of the sample to penetration by an indenter at the point of an instrument, which extends 0.1 in beyond the surface of the presser foot. The durometer reads 0 when the indenter is fully extended and 100 when it is pressed to a flat piece of plate glass; thus a high value indicates a hard sample. Coated fabric, to be hard, should give a reading of about 70.

HYDROSTATIC RESISTANCE

A measure of this characteristic is useful for coated fabric intended for rainproof applications. The Mullen test is for high water-pressure resistance. A sample is clamped between jaws with circular openings and water pressure is raised until the first drop of water passes through sample. Pressure at this point is expressed in MPa (psi; ASTM D 751).

The Suter test is used for medium or low water-pressure resistance (up to 14 kPa [2 psi]); it employs a rising water column placed over a test sample. Resistance is reported as the height, in cm, of the column at which pressure causes the first drop of water to penetrate the sample.

There is an impact spray test for low water-pressure resistance. A fabric sample is bombarded with a spray of water from a specified height, usually a column measuring 0.9 m (3 ft). The water that passes through a sample of a standard area within a specified time is accumulated in a 6 × 6 in blotter mounted behind fabric. The increase of weight, in mg, of the blotter is the spray penetration value.

LOW TEMPERATURE CRACKING

This guide is important with fabrics that must perform outdoors in cold climates, such as gas meter and regulator diaphragms. Test temperatures of –40°F and –67°F (–40°C and 19°C) are commonly used. In the bent loop and hammer test, a weight is dropped on a bent sample until it cracks. The test measure is given as the number of cycles to failure.

In the bar test, two samples measuring 1 in wide are threaded under a bar with ⅛ in diameter that forms the hinge between a center plate and a longer plate on each end. The end plates are lifted and then allowed to drop on the center plate, flexing the fabric samples around the bar until they crack. The measure of resistance is the number of cycles to failure.

In fold and roll tests, a 10 lb (4.5 kg) roller is rolled over a sample that is first folded in one direction and then folded in the opposite direction. The test is continued until the sample cracks, and the number of cycles to failure is used as the measure of resistance. Other tests include ASTM D 736, ASTM D 746, ASTM D 797, and ASTM D 1053.

MODULUS

A coating with a low modulus of elasticity is desirable in highly flexible coated fabric applications, such as rapidly pulsating diaphragms. A modulus can be measured satisfactorily only before the elastomer is applied to the base fabric. It is usually measured by the stress in MPa (psi) at any given elongation; most commonly used values are obtained at 300% elongation.

GAS PERMEABILITY

Permeability measurements of gases are important for coated fabrics used in life vests and other inflatable products. They are also vital for gas meter and regulator diaphragms. In the Cambridge permeameter test,samples are tested in an apparatus using hydrogen gas. Tests are usually conducted for 2 minutes, and the instrument converts readings into values expressed in $1/m^2/24$ h. For life vests and rafts, a reading of $5 1/m^2/24$ h is acceptable. Results can be converted into equivalent values for helium, carbon dioxide, or other gases by using factors recommended by the Bureau of Standards (ASTM D 815).

LIQUID PERMEABILITY

Liquid permeability measurements are significant for fabrics used as fuel containers or vapor traps, such as breather balloons. In a test, a specified amount of test liquid is placed in a cup or jar of specified dimensions; the fabric sample is then placed over the cup or jar and sealed. After allowing the sample to reach equilibrium, the cup or jar is weighed accurately and then inverted and allowed to stand for a specified period, after which it is weighed again. Weight loss is calculated generally in fl $oz/ft^2/24$ h.

pH VALUE

The pH value of a coating is useful to know in applications where the coating comes in contact with metal parts. Normally, the coating should have a neutral pH. The test method uses the fabric or coating cut into small pieces and placed in distilled water for 3 hours. The pH is then measured with a standard meter. A reading of 7 represents neutrality; < 7 shows acidity; > 7 shows alkalinity (Tables 10.24 and 10.25).

RESILIENCY

Resiliency can be an important index in many applications, such as rubber-coated offset printing blankets. The Bashore resiliometer test has a bob dropped vertically onto a fabric sample at the base of an instrument. The distance that the bob bounces back after impact is reported as a percentage of the original height from which it was dropped. It should be remembered that the resiliometer records the combined resiliency of the coating and base fabric. If a reading of the elastomer coating alone is wanted, then a fairly thick sample of elastomer coating material must be tested before the coating application.

STRETCH

Temporary stretch and permanent residual stretch are undesirable in a coated fabric that requires great accuracy, such as a metering diaphragm or an offset printing blanket. The test method uses a 10 in length (accurate to the nearest 0.01 in.) that is marked off on a 1×12 in sample. A clamp is

Acids (pH < 7)	Molarity	pH
Acetic	1 N	2.4
Acetic	0.1N	2.9
Acetic	0.01N	3.4
Alum	0.1N	3.2
Arsenious	Saturated	5.0
Benzoic	0.1N	3.0
Boric	0.1N	5.3
Carbonic	Saturated	3.8
Citric	0.1N	2.1
Formic	0.1N	2.3
Hydrochloric	1 N	0.1
Hydrochloric	0.1N	1.0
Hydrochloric	0.01N	2.0
Hydrocyanic	0.1N	5.1
Hydrogen Sulfide	0.1N	4.1
Lactic	0.1N	2.4
Malic	0.1N	2.2
Nitric	0.1N	1.0
Orthophosphoric	0.1N	1.5
Oxalic	0.1N	1.3
Succinic	0.1N	2.7
Salicylic	Saturated	2.4
Sulfuric	1 N	0.3
Sulfuric	0.1N	1.2
Sulfuric	0.01N	2.1

Bases (pH > 7)	Molarity	pH
Ammonia	1 N	11.6
Ammonia	0.1N	11.1
Ammonia	0.01N	10.6
Barbital Sodium	0.1N	9.4
Borax	0.01N	9.2
Calcium Carbonate	Saturated	9.4
Calcium Hydroxide	Saturated	12.4
Ferrous Hydroxide	Saturated	9.5
Lime	Saturated	12.4
Magnesia	Saturated	10.5
Potassium Acetate	0.1N	9.7
Potassium Bicarbonate	0.1N	8.2
Potassium Carbonate	0.1	11.5
Potassium Cyanide	0.1N	11.0
Potassium Hydroxide	1 N	14.0
Potassium Hydroxide	0.1N	13.0
Potassium Hydroxide	0.01N	12.0
Sodium Acetate	0.1N	8.9
Sodium Benzoate	0.1N	8.0
Sodium Bicarbonate	0.1N	8.4
Sodium Carbonate	0.1N	11.6
Sodium Hydroxide	1 N	14.0
Sodium Hydroxide	0.1N	13.0
Sodium Hydroxide	0.01N	12.0
Sodium Metasilicate	0.1N	12.6
Sodium Sesquicarbonate	0.1N	10.1
Trisodium Phosphate	0.1N	12.0

Table 10.24 Examples of acids and bases pH

Indicator	Acid Color	Low pH	High pH	Base Color
Methyl red	red	4.2	6.3	yellow
Azolitmin (litmus)	red	4.4	6.6	blue
Alizarin red S	yellow	4.6	6.0	red
Propyl red	red	4.6	6.6	yellow
Chlorophenol red	yellow	4.6	7.0	red
p-Nitrophenol	colorless	4.7	7.9	yellow
Bromophenol red	yellow	4.8	6.8	purple
Bromocresol purple	yellow	5.2	6.8	purple
p-Nitrophenol	colorless	5.4	6.6	yellow
Bromothymol blue	yellow	6.0	7.6	blue
Brilliant yellow	yellow	6.6	7.8	red
Phenol red	yellow	6.6	8.4	red
m-Nitrophenol	colorless	6.6	8.6	yellow
Neutral red	red	6.8	8.0	yellow
Rosolic acid	brown	6.9	8.0	red
Cresol red (2nd range)	yellow	7.0	8.8	red
a-Naphtholphthalein	brown	7.3	8.7	green
Methyl violet	yellow	0.0	1.6	blue
Cresol red (1st range)	red	0.0	1.8	yellow
Crystal violet	yellow	0.0	1.8	blue
Malachite green	yellow	0.2	1.8	blue-green
Methyl green	yellow	0.2	1.8	blue
Metanil yellow	red	1.2	2.4	yellow
m-Cresol purple (1st range)	red	1.2	2.8	yellow
Metacresol purple (1st range)	red	1.2	2.8	yellow
Thymol blue (1st range)	red	1.2	2.8	yellow
4-o-Tolylazo-o-toluidine	orange	1.4	2.8	yellow
Orange IV (Tropeolin 00)	red	1.4	3.2	yellow
2,6-Dinitrophenol	colorless	1.7	4.4	yellow
Benzyl orange	red	1.9	3.3	yellow
2,4-Dinitrophenol	colorless	2.0	4.7	yellow
Benzopurpurine 48	violet	2.2	4.2	red
p-Dimethylaminoazobenzene	red	2.9	4.0	yellow
Bromochlorophenol blue	yellow	3.0	4.6	purple
Bromophenol blue	yellow	3.0	4.6	blue
Congo red	blue	3.0	5.0	red
Methyl orange	red	3.1	4.4	yellow
Ethyl orange	red	3.4	4.8	yellow
Thymol blue (2nd range)	yellow	8.0	9.6	blue
Phenolphthalein	colorless	8.0	10.0	red
o-Cresolphthalein	colorless	8.2	9.8	red
Thymolphthalein	colorless	8.8	10.6	blue
Alizarin yellow GG	yellow	10.0	12.0	orange
b-Naphthol violet	yellow	10.0	12.0	violet
Alizarin yellow R	yellow	10.1	12.2	red
Nitramine	colorless	10.8	13.0	brown
Poirrier blue	blue	11.0	13.0	red
Resorcin yellow (Tropeolin 0)	yellow	11.1	12.7	orange
Benzene sulfonic acid	yellow	11.4	12.6	orange
Indigosulfonic acid	blue	11.4	13.0	yellow
Clayton yellow	yellow	12.2	13.2	amber

Table 10.25 Color indicators of acids and bases pH

applied to each end, and a 50 lb (22.5 kg) weight is suspended from one end for 10 minutes. The amount that the fabric stretches during testing is reported as temporary stretch and is expressed in a percentage of the original 10 in length. If a stretch remains after removing the weight, it is reported as permanent residual stress.

SWELLING

A test for swelling is required for any coated fabric that may come in contact with oils, solvents, or other liquids during use. There is a test method where oblong samples are immersed in a graduated cylinder containing alcohol to measure their volume. The samples are then dried and immersed in the test medium for a specified period under a specific temperature. Following exposure, the samples are removed and their volume is measured again. The increase in volume from swelling is recorded as percentage of the original volume.

Alcohol is used to measure the fabric volume because it is clear, does not cause swelling, and permits the sample to sink. Toluol is the most frequently used swelling medium because its swelling properties match those of many media. However, several other fluids are available for determining swelling characteristics.

Many brands of gasolines, oils, and other fluids have proprietary formula ingredients. Dust, fumes, and temperature, among other factors, may affect the composition of the fluids. Therefore, in critical applications such as fuel and brake systems, extensive life tests should be conducted with fabrics under the expected operating conditions.

If the coated fabric has to perform at elevated temperatures, then swelling tests are run at the expected temperatures. In some cases, the percentage increase in gauge or weight, rather than the change in volume, is measured.

TEAR STRENGTH

Fabric that will be stressed should be tested for tear strength. The Pendulum method (Elmendorf) test uses a 2.5 × 4 in fabric sample held between two clamps at the base of an Elmendorf instrument. The sample is nicked with a knife attachment, and then a pendulum is released and falls through the sample. The pendulum carries a circumferential scale that indicates the force required to tear the specimen. Scale readings can he multiplied by appropriate factors to give results in pounds or grams. Values are determined for both the warp and filling direction of the fabric (ASTM 751).

In the Tongue method (strip), a cut is made at the center of the 3 in edge of a 3 × 8 in sample. The two 1½ in ends are then placed in two jaws of a tester. The jaws are separated at a speed of 12 ipm. A recorder records tear as a high point on a chart. The average of five tests is reported as tear strength in lb or MPa. Values are determined for both warp and filling direction.

In the Trapezoid method, a trapezoidal sample, 6 in long on one side and 3 in long on the opposite parallel edge, is given a small cut at the center of the 3 in edge. The nonparallel edges are clamped into the jaws of a tester, and the sample is torn apart at a speed of 12 ipm. Using the recorder, a researcher notes the average of five high points as tear strength in MPa (lb).

TENSILE STRENGTH

The coated fabric manufacturer normally supplies a standard tensile strength value by the grab method in both warp and filling directions. The Grab method uses 4 × 6 in fabric samples that are inserted in 1 × 1 in jaws of a tester. The jaws, 3 in apart at the start, are separated at 12 ipm. Values are reported in pounds per inch. To obtain accurate results, extreme care must be used to obtain proper alignment of the yarns in the jaws (ASTM D 751).

In a cut-strip method, strips measuring 1 × 6 in are cut in both the warp and fill directions. In turn the yarns are carefully aligned in the dies. Samples are ruptured in a tester; values are recorded in lb-in (ASTM D 751).

THICKNESS

In many applications, it is essential to have uniform thickness over the entire surface of a coated fabric. Test methods using a gauge with 0.001 in graduations can be employed for thickness measurements. Since the gauge uses a deadweight rather than a spring gauge, great care must be taken to apply an absolutely straight load to obtain accurate readings.

WICKING

In some critical applications, it is essential to measure the air or gas leakage through pinholes or through internally exposed edges of vent holes into the fabric from where it bleeds to fabric edges. Coated fabric diaphragms performing in critical applications, such as air controls, must be proofed against wicking. In a wicking test, a coated fabric sample is clamped into a test jig immersed in water. Flange clamping pressure must be sufficient to prevent loss of gas between the sample and the flanges, but not so high as to impede the flow of gas through the edges.

After immersing the jig in water, gas pressure is applied to the jig, and pressure is gradually raised until bubbles are visible at the outer edge of the sample. Pressures are usually raised at increments of 10 psi up to a maximum of 100 psi. The pressure at which initial leakage occurs is recorded, and the next lowest pressure is defined as the antiwicking value, or the pressure up to which the fabric will perform without wicking.

WEATHERING

With any coated fabric that is to be used outdoors, it is very useful to have a guide to weathering properties before the fabric is placed in service. To obtain accelerated results, weatherometer equipment can be used to simulate rain and sunshine by use of water spray and carbon arc. Both physical tests and visual examination can be used to indicate the deterioration that occurs after a specified period of time. Military specifications frequently stipulate the minimum physical test values that must be met after a specified time of exposure. Comparative evaluations can provide an indication of the weather resistance of different fabrics (ASTM D750).

SOLVENT AND COATING

SOLVENT COMPOSITION IN COATING

The usual definition of solvents describes them as fluids having a maximum boiling point of 482°F (250°C) and able to dissolve other components of coatings, especially binders. They evaporate under drying conditions when paint films are formed. Solvents must not react with the coated product. The composition of coatings (paints, varnishes, etc.) is determined by application requirements, drying temperature, and drying time. Depending on the properties of paints and varnishes, different mixtures of solvents are added. Table 10.26 provides classifications and definitions of solvents (374).

The functions and properties of solvents in coatings are as follows:

1. Dissolve several components, especially binders.
2. Influence and control paint viscosity.
3. Wet pigments, thus influencing solubility, enabling hydrogen bonding by solvents, and preventing the separation of pigments.
4. Influence and control flow properties (e.g., butyl acetate, butanol, and glycol ethers).
5. Influence skin formation. The aim is to produce a homogeneous cure when, for example, the paint or varnish film hardens without the formation of a stable surface film during the drying period. The correct composition of the solvent will avoid trapping solvents under the surface film.
6. Influence the drying process, thus influencing the acceleration by low boilers or the production of a flawless surface by medium and high boilers (e.g., the chemical and physical drying processes).
7. Influence surface tension (e.g., increase by rapid evaporation of solvents).
8. Influence mechanical properties of paints (e.g., adhesion properties).
9. Influence blushing or blooming of paints by preventing the absorption of condensed water by various solvents (e.g., ethanol and glycol ether).
10. Influence gloss (e.g., improvement with high boilers).
11. Prevent defects particularly in varnish coating (e.g., background wetting).
12. Influence electrostatic properties (e.g., spray painting).
13. Influence defined surface properties (e.g., create structural changes).
14. Influence durability of paints and varnishes.
15. Influence product suitability (e.g., spraying and dipping lacquers that need to dry at room temperature).

Term	Definition
Aprotic/Protic	Aprotic solvents (also commonly called inert) have very little affinity for protons and are incapable to dissociating to give protons. Aprotic solvents are also called indifferent, non-dissociating, or non-ionizing. Protic solvents contain proton-donating groups.
Protogenic	An acidic solvent capable of donating protons.
Protophilic	A basic solvent able to combine with hydrogen ion or to act as a proton acceptor.
Acidic/Basic	Lewis acidity/basicity determines the solvent's ability to donate or accept a pair of electrons to form a coordinate bond with solute and/or between solvent molecules. A scale for this acid/base property was proposed by Gutman (DN and AN - donor and acceptor number, respectively) based on calorimetric determination. The complete proton transfer reaction with formation of protonated ions is determined by proton affinity, gas phase acidity, acid or base dissociation constants. Both concepts differ in terms of net chemical reaction.
Hydrogen-bonding	A bond involving a hydrogen atom, which is bound covalently with another atom, is referred to as hydrogen bonding. Two groups are involved: hydrogen donor (e.g., hydroxyl group) and hydrogen acceptor (e.g., carbonyl group).
Solvatochromism	Shift of UV/Vis absorption wavelength and intensity in the presence of solvents. A hypsochromic (blue) shift increases as solvent polarity increases. The shift in the red direction is called bathochromic.
Dielectric constant	A simple measure of solvent polarity (the electrostatic factor is a product of dielectric constant and dipole moment). The electrical conductivity of solvent indicates if there is a need to earth (or ground) the equipment which handles solvent to prevent static spark ignition. Admixtures affect solvent conductivity. These are most important in electronics industry.
Miscible	Solvents are usually miscible when their solubility parameters do not differ by more than 5 units. This general rule does not apply if one solvent is strongly polar.
Good solvent	Substances readily dissolve if the solubility parameters of solvent and solute are close (less than 6 units apart). This rule has some exceptions (for example, PVC is not soluble in toluene even though the difference of their solubility parameters is 2.5).
Θ solvent	The term relates to the temperature of any polymer/solvent pair at which chain expansion is exactly balanced by chain contraction. At this temperature, called Θ temperature chain dimensions are unperturbed by long-range interactions.
Reactivity	Solvent, according to this definition, should be a non-reactive medium but in some processes solvent will be consumed in the reaction to prevent its evaporation (and pollution). Solvents affect reactivity in two major ways: viscosity reduction and decreasing the barrier of Gibbs activation energy.
Hygroscopicity	Some solvents such as alcohols and glycols are hygroscopic and, as such, are unsuitable for certain applications which require a moisture-free environment or a predetermined freezing point. Solvents which are not hygroscopic may still contain moisture from dissolved water.

Table 10.26 Classifications and definitions of solvents

Term	Definition
Solvent strength	Solvent strength is used to establish required solvent concentration to form a clear solution and to estimate the diluting capabilities of pre-designed system. Two determined quantities are used for the purpose: Kauri butanol value and aniline point.
Solvent partition	Solvent partition is determined for three purposes: to estimate the potential for solvent removal from dilute solution by carbon black adsorption, to evaluate the partition of solute between water and solvent for the purpose of studying biological effects of solvents and solutes, and to design system for solvent extraction.
Volatility	Solvent volatility helps in estimation of the solvent evaporation rate at temperatures below its boiling point. The Knudsen, Henry, Cox, Antoine, and Clausius-Clapeyron equations are used to estimate the vapor pressure of a solvent over a liquid, its evaporation rate, and the composition of the atmosphere over the solvent. The boiling point of a solvent gives an indication of its evaporation rate but it is insufficient for its accurate estimation because of the influence of the molar enthalpy of evaporation.
Residue	This may refer to either the non-volatile residue or the potential for residual solvent left after processing. The former can be estimated from the solvent specification, the later is determined by system and technology design.
Carcinogenic	Solvents may belong to a group of carcinogenic substances. Several groups of solvents have representatives in this category (see listings in Section 3.3)
Mutagenic	Mutagenic substance causes genetic alterations, such as genetic mutation or a change to the structure and number of chromosomes (mutagens listed in Section 3.3).
Impairing reproduction	Several solvents in the glycols and formamides groups are considered to impair fertility.
Toxicity	LD50 and LC50 give toxicity in mg per kg of body weight or ppm, respectively. Threshold limit values place a limit on permissible concentration of solvent vapors in the work place. Also "immediate-danger-to-life" and "short-term-exposure-limits" are specified for solvents. Odor threshold values have limited use in evaluating the potential danger to solvent exposure.
Flammable	Several data are used to evaluate the dangers of solvent explosion and flammability. Flash point and autoignition temperature are used to determine a solvent's flammability and its potential for ignition. The flash points for hydrocarbons correlate with their initial boiling points. Lower and upper explosive limits determine the safe ranges of solvent concentration.
Combustible	The net heat of combustion and the calorific value help to estimate the potential energy which can be recovered from burning used solvents. In addition, the composition of the combustion products is considered to evaluate potential corrosiveness and the effect on the environment.
Ozone depleter	Ozone depletion potential is the value relative to that of CFC-11. It represents the amount of ozone destroyed by the emission of gas over its entire atmospheric life-time. Photochemical ozone creation potential is a relative value to that of ethene to form ozone in an urban environment. Numerous solvents belong to both groups.
Biodegradability	Several methods are used to express biodegradability. These include biodegradation half-life, biological oxygen demand, chemical and theoretical oxygen demand.
Cost	Cost of solvent is a key factor in solvent selection.

Table 10.26 Classifications and definitions of solvents *(continued)*

Term	Definition
Solvent	A substance that dissolves other material(s) to form solution. Common solvents are liquid at room temperature but can be solid (ionic solvents) or gas (carbon dioxide). Solvents are differentiated from plasticizers by limiting their boiling point to a maximum of 250°C. To differentiate solvents from monomers and other reactive materials - a solvent is considered to be non-reactive.
Polarity	Polarity is the ability to form two opposite centers in the molecule. The concept is used in solvents to describe their dissolving capabilities or the interactive forces between solvent and solute. Because it depends on dipole moment, hydrogen bonding, entropy, and enthalpy, it is a composite property without a physical definition. The dipole moment has the greatest influence on polar properties of solvents. Highly symmetrical molecules (e.g. benzene) and aliphatic hydrocarbons (e.g. hexane) have no dipole moment and are considered non-polar. Dimethyl sulfoxide, ketones, esters, alcohol are examples of compounds having dipole moments (from high to medium, sequentially) and they are polar, medium polar, and dipolar liquids.
Polarizability	The molecules of some solvents are electrically neutral but dipoles can be induced by external electromagnetic field.
Normal	A normal solvent does not undergo chemical associations (e.g. the formation of complexes between its molecules).

Table 10.26 Classifications and definitions of solvents *(continued)*

In addition to their effect on performance and properties solvents interact with other components in paints and varnishes in significant ways. Interactions between binders and solvents in paints and varnishes are very important. With the aid of solubility parameters, solvents or mixtures of solvents that produce the required properties may be selected. The influence extends to the dissolving of binders, the reduction of paint viscosity, pigment wetting, and so on.

Optimized dissolving of binders can be achieved by selecting the appropriate solvent mixtures, in which the density approaches that of the binder solubility range. Nevertheless, the selection of an optimized solvent mixture is difficult because there are conflicting requirements and outcomes. In one aspect, the chemical nature of the solvents should be similar to those of the binder to improve the flow. In another aspect, the solubility and hydrogen-bonding abilities of the solvents should be at the edge of the binder solubility range, because this results in rapid drying with a low retention of solvents.

SOLVENT AND SOLVENT-FREE COATINGS

Solvent composition is an important aspect in classifying paints and varnishes. A list of the main groups of these coatings follows:

1. *Solvent products.* These products contain solvents of various mixtures, types, and concentrations depending on the properties desired (e.g., application method, surface film, or skin formation). Solvents are normally the main components of these products.

2. *Solvent-reduced products.* These contain solvents in lower concentrations compared to conventional products and thus a higher content of solids. The main groups of solvent-reduced paints are medium solid contents (55% to 65% solids) and high solid contents (60% to 80% solids).
3. *Waterborne coatings.* This group contains deionized water as a dispersing agent. Normally cosolvents are added (up to 25%). The term "waterborne coatings" is mainly applied to industrial coating materials, which differ from silicate colors, wood-preservative varnishes, and emulsion paints.
4. *Solvent-free products.* The products of this group are produced and applied without organic solvents. They include powder coatings, radiation curing systems, and solvent-free water coating (without cosolvents).

The use of paints and varnishes containing high solvent concentrations is becoming less common; solvent-reduced products, waterborne coatings, and solvent-free products are applied more often. Whereas environmental and health-related concerns call for the reduction of solvents in paints and varnish products, qualitative aspects still demand the use of solvents in some fields of application.

Solvent-reduced products have achieved the same qualitative properties as solvent-containing products (e.g., application properties, periods of guarantee, limited costs, loading capacities, surface properties). Solvent-reduced or solvent-free varnish products have been produced with high quality (e.g., durable, good application properties) and limited costs. In some fields of application, such as waterborne coatings, high solids in varnish coatings of vehicles, a lot of developmental work has been done.

Additionally, other components in varnishes apart from solvents or modifications of application techniques can improve the properties of solvent-reduced paint and varnish products. Nevertheless, a wide rage of quality exists in paints and varnish products that are offered commercially and, in some cases (e.g., film-forming processes, processability, corrosion protection, purification, special applications, or wood preservation), solvent-based products are still preferred. In the field of wood preservation especially, solvent-based products (alkyd resins) are used because of some technical advantages (e.g., more solid surfaces), but alternative high solid systems are available.

EMISSION

Solvents are usually the most significant emission products coming from building materials and interior furnishings. All painted products are potential sources of emission. Even the so-called "bio" paints or natural paints emit various substances. Examples of these include mineral varnishes, natural oils, and even synthetic terpene-like compounds. Depending on the products and the components that make them up, the following various parameters can determine the emissions and behavior of solvents in ambient air from paints and varnishes (374):

1. Emission of solvents during the film-formation stage. The emission rate is directly proportional to the VOC concentrations in the product and inversely proportional to the film thickness (first order of kinetics). When the film has completely formed, the emission is controlled by a diffusion process, and the emission rate is now inversely proportional to the square of the film thickness.
2. Application of the paints and varnishes, methods of application of the paint or varnish (e.g., speed of application of the paint).
3. Characteristics of solvents in paints and varnishes (e.g., volatility or the boiling point, dynamic characteristics of evaporation and concentrations). Substances that have a low boiling point evaporate fast, mostly during application, and cause a rapid skin formation. Thus the risk of exposure is mainly with the painters. Medium boilers allow the surface to remain open for a while (evaporation of volatile products). The evaporation of substances with a high boiling point is slow, taking several weeks or months after application; as a result, a building's occupants may be exposed to the substances.
4. Characteristics of other compounds in paints and varnishes (e.g., relationship of binders to solvents, possible reactions between solvents and other compounds).
5. Characteristics of surfaces that have been painted (e.g., area, structure of surface).
6. Characteristics of emission processes (e.g., diffusion), dynamics of emissions (constant of evaporation), interrelations (e.g., diffusion and back diffusion).

The quantitative assessment of emission processes can be described with various models. The usefulness of these models differs. Some models describe these processes very well, as proven by various experiments or measurements (e.g., test chambers). Basic equations that describe emission processes are shown in Table 10.27. An emission of solvents from a varnish system occurs in the course of a varnish's life cycle in several different locations, as shown in Figure 10.21.

The emission processes of solvents from paints and varnishes can be divided into two phases:

1. Emissions during application of paints, which deals with complex interrelations dependent on various parameters
2. Emissions after the application process, the course of which is governed by complex emission processes that are dependent on various parameters (e.g., film formation, surface area)

Most solvent products, especially organic solvents and some additives, emitted from paints and varnishes are VOCs. The largest components of VOCs are solvents (e.g., aliphatic and aromatic hydrocarbons, alcohols, amines, acids, aldehydes, esters, ketones, and terpenes). The definition of VOCs varies. A standard definition is published by European Committee for Standardization (CEN): VOCs are any organic liquids and/or solids that evaporate spontaneously at the prevailing

1. At the beginning of the application process (t=0) the mass of VOC changes positively ($v_{ST}d_sc_w$), on the other hand VOC evaporates (first order of kinetics).
   $dm_w/dt = v_{ST}d_sc_w - k_1m_w/d_s$
2. In the ambient air the mass of VOC increases because of the evaporation out of the wall and decreases according to the ventilation rates.
   $dm_L/dt = k_1m_w/d_s - k_2m_L$
3. If the connected differential equations are solved and integrated (from t=0 until the end of application $t=A/v_{ST}$), the following equations are received:
   $m_w(t) = v_{ST}d_s^2c_w/k_1(1-\exp(-k_1t/d_s))$
   and
   $m_L(t) = ((1-\exp(Bt)/B - (1-\exp(k_2t)/k_2)v_{ST}d_sc_w\exp(-k_2t))$
   with
   $B = k_2 - k_1/d_2$
4. After finishing application only evaporation is relevant (equation is simplified):
   $dm_w/dt = - k_1m_W/d_s$
5. The course of VOC in the ambient air does not change (equation corresponds to equation):
   $dm_L/dt = k_1m_w/d_s - k_2m_L$
6. The solution of these differential equations describes the quantities of VOC in the wall (equation and the course of VOC in the ambient air):
   $m_w(t) = m_{w.AE}\exp(-k_1(t-t_{AE})/d_s)$
   $m_L(t) = ((k_1m_{w.AE}/d_s)(\exp(B(t- t_{AE}))-1)/B + m_{L.AE})\ \exp(-k_2(t-t_{AE}))$
   with
   $B = k_2 - k_1/d_s$

where:

A	area of the wall
a	coating thickness
B	fraction of binder
c_W	VOC-concentration in the wall
c_L	VOC-concentration in the ambient air
D	density
d_s	thickness of the layer of the paint application (=a/D)
k_1	constant of evaporation
k_2	ventilation rate of the indoor air
m_L	mass of VOC in the indoor air
$m_{L.AE}$	m_L at the end of the application
m_w	mass of VOC in the wall
$m_{W.AE}$	m_W at the end of the application
R_M	VOC-content in the dispersion
t	time
V	volume of the indoor air
v_{ST}	spreading velocity

Table 10.27 Examples of basic calculations of VOC-emissions during applications of emulsion paints

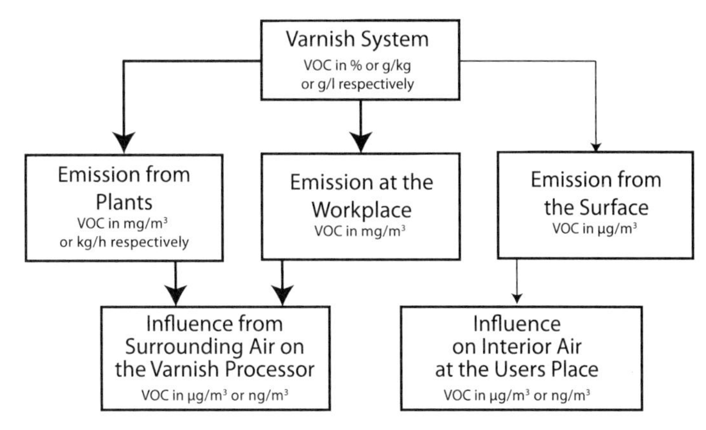

Figure 10.21 Emission of VOCs in the life cycle of a varnish.

temperature and pressure of the atmosphere. VOC content (VOCC) is defined as the mass of the VOCs in a coating material, as determined under specific conditions.

CLEAN AIR ACT

Almost all solvents are VOCs and hazardous air pollutants (HAPs), and their evaporation creates environmental problems that have become the focus of many domestic and international regulations and initiatives. A VOC solvent is defined by the Environmental Protection Agency (EPA) as any compound of carbon, excluding carbon monoxide, carbon dioxide, carbonic acid, metallic carbides or carbonates, and ammonium carbonate, which is emitted or evaporated into the atmosphere.

In recent years, the US Congress and the EPA have recognized the need to address environmental problems from a more holistic approach, considering multimedia and innovative environmental management strategies. Thus various Clean Air Act programs and project initiatives, which have proven to be extremely successful, have emerged. Many of them encourage voluntary industry participation and do not take the old command-and-control approach. Details of the Clean Air Act are reviewed in chapter 26.

SOLVENT SUBSTITUTION

In the past few decades, substitution of solvents by safer products and processes has been occurring. Effective utilization of supercritical fluids has been developed for high-value-added differentiated products.

A new wave of second-generation supercritical technologies has started to emerge, creating new roles for dense gases. "Supercritical" refers to the state of matter in which the temperature and pressure of a single-component fluid are above the critical point at which the phase boundaries diminish. A portfolio of chemical and physical operations carried out in the vicinity of this region defines supercritical fluid technology (SFT).

The pressure–temperature–volume (PVT) behavior of a substance can be best depicted by pressure—-temperature and pressure–density (volume) projections, as shown in Figure 10.22. The pressure–temperature diagram identifies the supercritical fluid region, alternative separation techniques that involve phase transition including the associated phase boundaries, and the fact that an isotherm below critical (AB; Fig. 10.22) involves phase transition while one above (A′B′) does not. Pressure–density projections illustrate tunability of the solvent density at supercritical conditions (A′B′), and again the continuity of the isotherm that does not encounter any phase transition. The aforementioned behavior is for a pure-component solvent. The critical properties of various organic and inorganic substances are shown in Table 10.28.

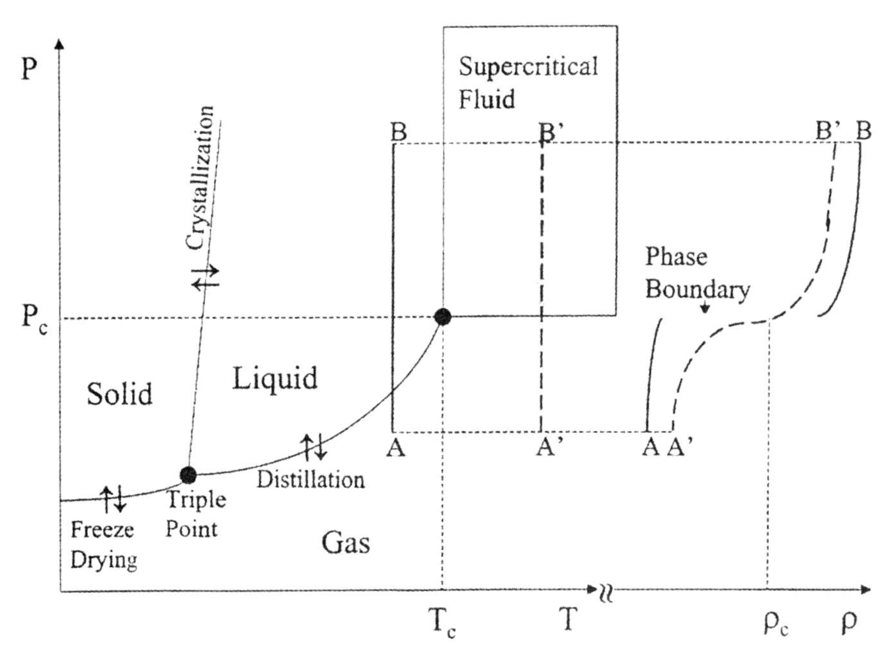

Figure 10.22 Pressure-temperature and pressure-density behavior of matter.

Solvents	Critical temperature, °C	Critical pressure, atm
Critical conditions for various inorganic supercritical solvents		
Ammonia	132.5	112.5
Carbon dioxide	31.0	72.9
Carbonyl sulfide	104.8	65.0
Nitric oxide	-93.0	64.0
Nitrous oxide	36.5	71.7
Chlorotrifluoro silane	34.5	34.2
Silane	-3.46	47.8
Xenon	16.6	58.0
Water	374.1	218.3
Critical conditions for various organic supercritical solvents		
Acetone	235.5	47.0
Ethane	32.3	48.2
Ethanol	243.0	63.0
Ethylene	9.3	49.7
Propane	96.7	41.9
Propylene	91.9	45.6
Cyclohexane	280.3	40.2
Isopropanol	235.2	47.0
Benzene	289.0	48.3
Toluene	318.6	40.6
p-Xylene	343.1	34.7
Chlorofluoromethane	28.9	38.7
Trichlorofluoromethane	198.1	43.5

Table 10.28 Critical properties of solvents

Chapter 11

Casting

INTRODUCTION

By definition, casting involves the formation of an object by pouring a liquid plastic into an open mold or surface where it completes its solidification. Either liquid thermoplastic (TP) or thermoset (TS) plastic is used. TP hardens after it is poured. The TS plastic chemically reacts and cures to form a rigid product (chapter 1). The choice of casting material, the type of mold, and the method of fabrication often depend on the application. Although production is rarely automated, it may be when the economic benefits are considerable.

Casting may be used to fabricate different-shaped products, rods, tubes, and so on, by pouring a liquid solution into an open or closed mold, where it hardens into a solid product. Film and sheeting are also made by casting directly into a flat open mold; by casting onto a wheel, continuously moving turntable, or conveyor belt; or by precipitation in a chemical bath. Various ornamental or utilitarian objects are often embedded in the plastic. The process provides a means to easily incorporate different product requirements such as coloring and texturing. The plastic mixture may contain pigments, fillers, plasticizer, and other chemical additives. It can enhance aesthetic quality of the product and increase its strength (chapter 15).

One essential difference between casting and molding processes is that pressure need not be used in casting (although large-volume, complex parts can be made by either pressure or vacuum-casting methods). Casting depends on gravity and heat that is either self-generated (such as by using a catalyst) or externally applied to solidify the mass. Another difference is that, in casting, the starting material is usually in liquid rather than solid form (e.g., pellets, granules, flakes, or powder). There is also a difference in that the liquid is often a monomer rather than a polymer or plastic, as is used in most molding compounds (chapter 1).

Casting has the advantage of using inexpensive equipment, but it is a relatively slow process. Casting to fabricate products that have complex shapes or certain other properties requires skill, especially for large castings, and the methods used constitute an art.

PLASTIC

Generally plastics that are free-flowing and have low surface tensions with low viscosities are used for castings of intricate shapes and finely detailed design. Low-viscosity plastics are also more suitable for producing bubble-free castings. High-viscosity systems usually produce castings with better physical properties than do low-viscosity plastics. High-viscosity plastics require close attention to handling procedures because they are usually more difficult to process.

Most plastics suitable for castings are two-component systems. A specified amount of hardener or accelerator is added to the plastic. It is important to ensure that thorough mixing takes place to maximize performance. Prior to pouring the compound into a mold, it is usually coated with a mold-release agent. For fabricating certain products air is removed, usually by using a vacuum system prior to the solidification of the plastic.

Depending on the plastic to be cast, solidification takes place at either room temperature or elevated temperatures. With room temperature systems chemical reaction occurs with the liberation of heat. The rate of heat dissipation can influence the performance and aesthetic characteristics of the hardened product. In thin sections, where a large area is exposed in relation to the total volume of the plastic, the heat of the exothermic reaction is dissipated rapidly and the temperature of casting is not very high. Thus, thin sections can be cast at room temperature with no danger of cracking. When the rate of heat dissipation is excessive, application of heat may be necessary to properly control the cure rate.

When the loss of heat is excessive, such as with thick castings, proper application or rate of heating may be necessary to accomplish the cure without any damage. The compound ingredients and their proportions can control the rate of heat development. Heat sinks (or heat pipes) can be used for the absorption or transfer of heat away from a critical element or part.

Air entrapment is a phenomenon wherein air is occulated in a plastic or composite system, giving rise to blisters, bubbles, and/or voids that are usually not desired. It can occur during any fabricating method. The bubbles could be due to air alone or moisture caused by improper drying of the plastic material, volatile compounding agents, plastic degradation, or the use of contaminated regrind material. So the first step in resolving this problem is to determine which of these problems exists. A logical troubleshooting approach can be used.

During casting, air bubbles can be present. Sometimes they are invisible and other times they are visible but not materially damaging. During casting, damaging or unwanted air bubbles could be present due to material preparation and/or during processing. Methods for their removal exist. Air is present in the plastic with its hardener, catalyst, or other additives and reinforcements. The air exists in the compounds when they are mixed and poured into the mold. The number of air bubbles that form depends on the viscosity and surface tension of the plastic-hardener system, the solubility of air in the system, and the characteristics of the mold surface. Although fillers extend the plastic and lower bulk cost, they tend to retain air. Deaerating systems, including vacuum systems, are used when unwanted air develops during mixing.

Air removal problems due to fabricating can be avoided by proper design of the mold. Elimination of sharp corners and provisions of an adequate venting system can facilitate removal of the entrapped air (chapter 17).

PROCESSES

Different casting processes are used. They tend to overlap and could be identified by other processing methods. An example is liquid injection molding (LIM), which can be identified as injection molding (chapter 4) or reaction injection molding (RIM; chapter 12). Many decades ago the RIM process was initially called LIM.

Different methods using different plastics are used. In most cases, casting corresponds to the pouring of the liquid plastic into a mold (gravity or atmospheric pressure casting; Fig. 11.1). Because of the low stresses involved, lightly built open molds or two-piece molds may be used. The relatively high viscosity of the monomer or reactants, and the need to vent the cavity, often calls for mold designs reminiscent of metal die casting (bottom feed runners, vents, etc.).

In some cases, the chemical reaction that takes place during a casting process converts a low molecular weight monomer into a high molecular weight TP (chapters 1 and 2). The most common examples are acrylics and nylons. In other cases, polymerization and cross-linking take place simultaneously in the casting process, leading to TSs. Examples include polyurethane resins (PUR), unsaturated polyester plastics (UP), epoxy plastics (EP), and silicone plastics (ST).

As reviewed with room temperature exothermic heat curing systems, additives and/or promoters are used to provide the necessary heat through chemical reactions. This reaction has to be controlled so that overheating will not occur in large parts of the product.

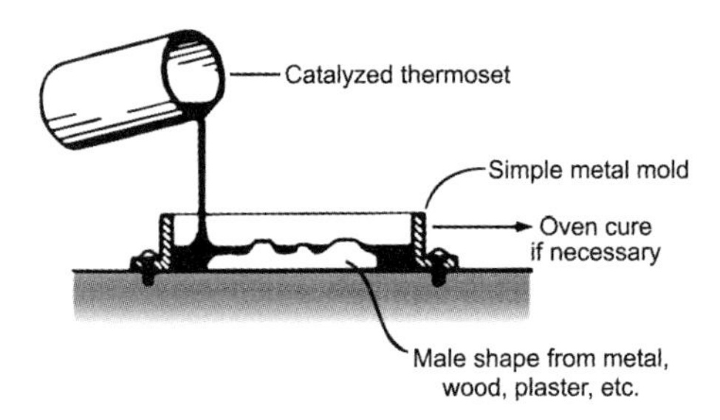

Figure 11.1 Example of the liquid casting process.

In addition to the conventional liquid-pouring casting process, other methods are used, including investment casting, which dates back centuries. The early Egyptians developed investment casting to make jewelry, dipping sculpture wax into a ceramic slurry, then drying, and heating to remove the wax. In turn the ceramic cavity received molten metal to form the desired finished part. This technique continued to be used with modifications that initially led to the casting of different materials, including plastics. Other systems evolved, such as the lost wax or soluble core wax. Later low-melting eutectic alloys were used, providing a means to produce high-production complex castings (chapter 15).

The centrifugal casting process, also called centrifugal molding, is a method of forming plastic in which a dry or liquid plastic is placed in a rotating mold such as a pipe (chapter 13). As it rotates around a single axis, heat is applied to the mold. The centrifugal force induced forces the molten plastic to conform to the configuration of the inside mold cavity. This method is different from rotational molding since the mold rotates only around one axis.

Products such as tubing, pipes, and tanks (excluding end caps), which have a circular-cylindrical shape, can be made of unreinforced or glass fiber–reinforced plastic (GRP) by the centrifugal casting process. In the case of discontinuous fiber reinforcement, a mix of chopped fibers and precatalyzed liquid plastic is dispensed along the axis of a rotating cylindrical tool. As the material falls on the inside of the tool surface, it is entrained and centrifugal forces help compact it into a uniform layer, also keeping it in place during cure. Successive passes along the length of the cylinder can build up thickness. The final inner surface, although not as good as the outer surface, is reasonably smooth. Features can be formed in the outer surface (flanges, threads, ribs, etc.) if the mold is made of suitable sections to allow the extraction of the finished product. The process can be modified to allow continuous pipe production.

Modifications of the casting process are used to meet different product and/or cost requirements. In an extension of the process, large pipes have been made that involve the successive formation of an external reinforced plastic (RP) layer, a rigid foam core layer, and an internal RP layer. Such pipes have outstanding insulation and crushing properties, in addition to pressure resistance.

There is a modified centrifugal casting process that produces continuous filament-reinforced TP pipes/tubes with precise fiber placement and with smooth internal and external surfaces. TPs such as nylon and polypropylene can be reinforced with fibers such as glass and carbon. Products such as automotive drive shafts and bearings have been fabricated with fiber volumes up to 60 wt%. These tubes have very low rotational imbalances and tight tolerance of wall thickness.

This process, called TER-centrifuging (H. Schurmann, Technische University, Darmstadt, Germany), starts by winding dry reinforcing fibers around a TP tube that can be made by extrusion or injection molding. The fibers can be arranged to meet a specific load requirement. The tube and fibers are then loaded into a casting mold, rotated at a controlled rate, and heated. As the molten plastic tube rotates, plastic impregnates the fibers.

There is also dip casting, which is also called dip coating or dip molding. It is a process of submerging a hot molded shape, usually metal, into a fluid plastic. After removal and cooling, the product around the mold is removed from the mold.

Slush casting, also called slush molding or cast molding, is extensively used. It is a method in which TPs in a liquid form are poured into a hot mold, stationary or moving, where a viscous skin forms. The excess slush is drained off, the mold is cooled, and the molding stripped out. This method is used to produce rain or snow boats, auto instrument panels, overshoes, corrugated and non-corrugated complex tubes, caps, and other shapes.

With solvent casting, a plastic compounded with its constituents (solvent, stabilizers, additives, plasticizers, etc.) is carefully prepared at a certain rate of mixing. These soluble plastics are poured into a mold or onto a moving belt (to form film) where heat is applied using heat control zones to prevent formation of blisters. The rate of solvent evaporation is inversely proportional to the square of the thickness. To reduce cost and meet regulations, solvent recovery systems are used that have explosive-proof hazard safety capabilities. There are also systems that use water-based solvent solutions such as polyvinyl alcohol plastic.

Spin casting can use plastic molds, such as silicone, to produce close-tolerance, highly cost-effective, limited production in a variety of materials. The process uses easily adjustable centrifugal force to inject liquid TS plastics into a circular disk-shaped elastomeric molds under pressure, completely and rapidly filling the mold cavities.

Vacuum casting is the process used in the different casting processes in which a vacuum is required to withdraw air from the different casting materials before they harden or cure, and it is designed to prevent air bubble defects in the finished product.

A simple nonmechanical version of RIM (chapter 12) or LIM is foam casting. Foaming components are poured into a mold cavity that is usually heated (chapter 8).

LIM involves the proportioning, mixing, and dispensing of liquid components and injecting the resulting mixture directly into a mold cavity that has been clamped under pressure. In this casting process, the liquid is injected under pressure that is far less than that used in conventional injection molding (chapter 4). A simplified view of this casting process is shown in Figure 11.2. For more precision, mixing equipment is available, such as that shown in the schematic in Figure 11.3.

A similar process is RIM. The LIM process involves mechanical proportioning, mixing, and dispensing of two liquid plastic formulations, and then the mixture is directed into a closed mold. It can be used for encapsulating electrical and electronic devices, decorative ornaments, medical devices, auto parts, and so on. The difference is that RIM uses a high-pressure impingement mixer instead of mechanical mixing. Flushing the mixture at the end of a run is easily handled automatically. Plastics used include silicones, acrylics, and so on. To avoid liquid injection hardware from becoming plugged with plastics, a spring-loaded, pin-type nozzle may be used. The spring loading allows setting the pressure so that it is higher than the pressure inside the extruder barrel, thus keeping the port clean and open.

Different foundry casting techniques are used. Included are plastic-based binders mixed with sand. Various types of molds and cores are produced, including no-bake or cold-box, hot-box, shell, and oven-cured. The usual binders are phenolic, furan, and TS polyester.

There is the foundry shell casting, also called dry-mix molding. It is a type of process used in the foundry industry, in which a mixture of sand and plastic (phenolic, TS polyester, etc.) is placed

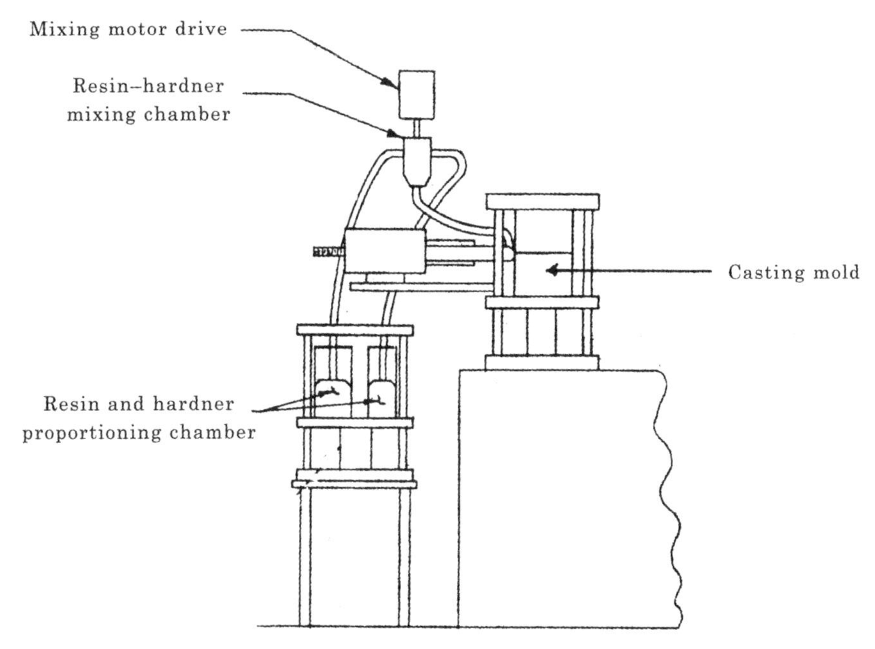

Figure 11.2 Example of a LIM casting process.

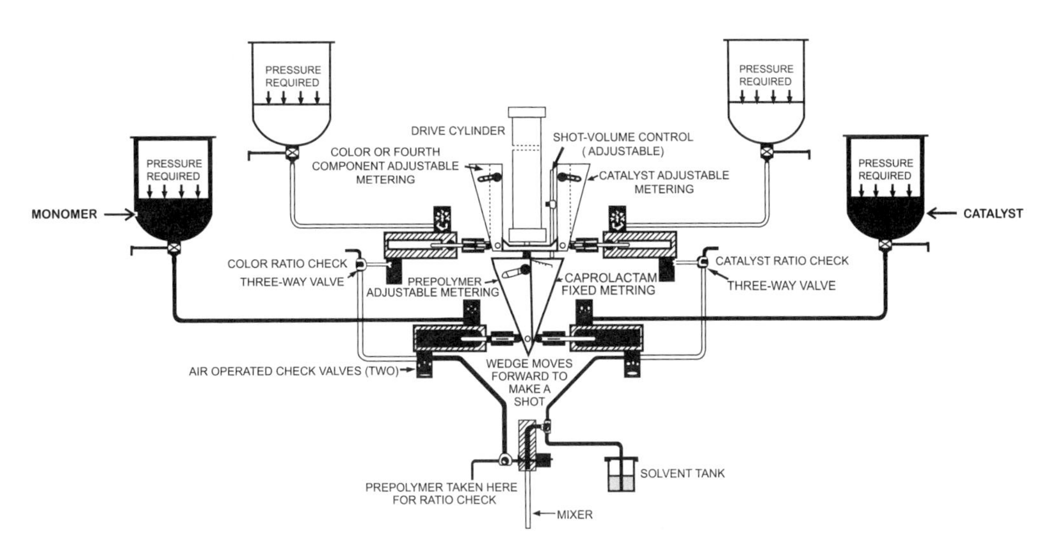

Figure 11.3 Example of more accurate mixing of components for liquid injection casting.

onto a preheated metal pattern (producing half a mold), causing the plastic to flow and build a thin shell over the pattern. Liquid plastic precoated sand is also used. After a short cure time at high temperature, the mold is stripped from its pattern and combined with a similar half produced by the same technique. The finished mold is then ready to receive the molten metal. Blowing a liquid plastic/sand mix in a core-box also produces shell molds.

Different materials are impregnated with different plastics to provide increased performances and/or decorations. It includes application as a matrix to reinforcing fiber producing exceptionally high strength structures, saturating cement/concrete or wood to improve strength and extend environmental endurance, and filling metals that are slightly porous to seal them. The degree of impregnation or saturation depends on variables such as the process used, which may include casting, coating, extrusion, tower drying, and so on, with or without a vacuum in the substrate.

The trickle impregnation process is related to TS plastic casting, potting, and encapsulation. It also uses a low viscosity liquid reactive plastic to provide the trickle impregnation. As an example, the catalyzed plastic drips onto an electrical transformer coil. Capillary action draws the liquid into the coil's openings at a rate slow enough to enable air to escape as it is displaced by the liquid. When fully impregnated, the part is exposed to heat to cure the plastic.

CASTING OF ACRYLIC

INTRODUCTION

Casting acrylic dates back to the 1930s. (Literature has been rather extensive on this subject; only a summary is provided in this section.) Individual casting is used to produce excellent optical properties. Sheeting is also made by continuous casting (sometimes called extrusion casting); a monomer–plastic–catalyst mixture is fed onto a stainless steel belt on which polymerization is completed. Continuous casting is not as optically clear as cast sheeting; parallelism and flatness of surfaces, as well as optically uniform density, is not nearly as uniform.

Acrylic castings usually consist of polymethyl methacrylate or copolymers of this ester as the major component, with small amounts of other monomers to modify the properties (chapter 2). Incorporating acrylates or higher methacrylates, for example, lowers the heat deflection temperature and hardness and improves thermoformability and solvent cementing capability, with some loss in resistance to weathering. Dimethacrylates or other cross-linking monomers increase the resistance to solvents and moisture.

The acrylic castings are made by pouring the monomers or partially polymerized syrups into suitably designed molds and heating to complete the polymerization. A large reduction in volume, about 21%, takes place during the cure. The reaction also is accompanied by the liberation of a substantial amount of heat, 13.8 kilocalories/mole. At conversions above 20%, the polymerization becomes accelerated, and the rate rises rapidly until gelation occurs at about 85% conversion. Thereafter, the reaction slows down and a post-cure may be needed to complete the polymerization.

During the accelerated phase, the rapid increase in viscosity and liberation of heat can raise the internal temperature and elevate the reaction rate unless measures are taken to dissipate the heat; otherwise, in extreme cases, a violent runaway polymerization can occur. During this cure cycle the effects of shrinkage and acceleration can be controlled by interrupting the polymerization to form syrup containing 25% to 50% polymer. Syrups can be stored safely with little change until they are needed, and the amounts of shrinkage and heat production drop during the second stage of cure in accordance with the polymer content. Using syrups also shortens the time in the mold, decreases the tendency to leakage from the molds, and greatly decreases the chance of dangerous runaways.

For acrylic products needing high optical quality, syrups are produced by careful heating with constant stirring of monomer containing a small amount (0.02% to 0.1%) of a soluble free-radical initiator (peroxides, etc.) until a molasses-like consistency is achieved. To make syrups of lower viscosity small amounts (0.1% to 0.4%) of sulfur containing chain transfer agents, or modifiers, such as lauryl mercaptan or mercaptoesters, are charged with the monomers.

Applications of acrylic sheet include the fabrication of signs and displays, breakage-resistant and security glazing, enclosures, skylights, fascia and panels, display containers, sunscreens, furniture, balustrades, separators, and aircraft glazing.

Manufacturers of acrylic castings must recognize the toxicity, flammability, and explosive potential of methyl methacrylate and peroxides. Sources of ignition must be kept away from these materials, and adequate, reliable ventilation and means of removing vapors must be provided in storage and processing areas. The monomer is moderately toxic in the liquid and vapor state. It irritates the eyes, produces sensitization of the skin, and causes toxic or allergenic reactions in susceptible persons. Personnel handling the monomer or involved in cleanup of spills must wear goggles and impervious gloves and maintain strict personal cleanliness. Material safety data sheets and other information on dealing with methyl methacrylate are available from suppliers.

Acrylics are combustible plastics, and the fire precautions normally used with other combustibles must be observed in handling, storing, and using them. The fire hazards of acrylic installations can be kept within acceptable levels by complying with building codes, applicable Underwriters Laboratories standards, and the established principles of fire safety.

CASTING SHEET

Extensive use is made of casting sheets. Cast sheet is made in a batch process within a mold or cell or is produced continuously between stainless steel belts. Basically the processing cells consist of two pieces of polished (or tempered) plate glass slightly larger in area than the finished sheet is to be. The cell is held together by spring clips that respond to the contraction of the acrylic material during the cure.

The plates are separated by a flexible gasket of plasticized polyvinyl chloride (PVC) tubing that controls the thickness of the product. The cell is prepared by fitting the gasket between the plates and clamping, while leaving a corner open. The cell is tilted slightly from the horizontal and filled with a weighed amount of catalyzed syrup containing any required plasticizers, modifiers, release agents, colorants, ultraviolet absorbers, flame retardants, and so on. The rest of the gasket is then set in place and clamped. The filled cell is returned to a horizontal position and moved into an oven for cure.

A thin sheet (less than 0.5 in) is cured in a forced-draft oven using a programmed temperature cycle starting at about 45°C and ending near 900°C. The cycle is 12 to 16 hours for a 0.125 in sheet and considerably longer for thicker sheets. Thicker sheets are best made in an oil or water bath or in an autoclave. Because of the poor thermal conductivity of air, the heat of polymerization in a forced draft oven can drive the temperature within the mass far above the boiling point (100°C) of acrylic.

Bubbles of monomer may form and become trapped within the plastic to mar its clarity and appearance. Operation in a liquid bath dissipates the heat more efficiently to help prevent local overheating and bubbling. Operation in an autoclave under a pressure up to 100 psi also keeps the monomer from boiling and gives clear undistorted products. Autoclave processes are practical and usually necessary for producing irregular shapes and thick sheets.

After final curing, the cell-cast sheet is cooled in the molds, stripped, and trimmed to size. The sheet is supplied in sizes up to 120 by 144 in and thicknesses from 0.03 to 4.25 in. A wide range of colors in clear, translucent, and opaque products is available in standard and custom grades. The sheets come unmasked or with easily strippable Kraft paper or polyethylene masking on one or both sides to protect the plastic from damage and scratching during shipping, storing, and cutting.

In the continuous casting process viscous syrup is cured between two highly polished moving stainless steel belts. The distance between the belts determines the thickness of the sheets. Their width is controlled by inserting flexible gaskets between the belts and is limited only by the width of the belts. Continuous casting is less versatile than cell casting and is limited to relatively thin (up to 0.375 in) sheets.

An important advantage of the method is the elimination of the severe problems of handling and breakage of large sheets of costly glass that make up the cells. Cell-cast sheet has superior optical properties and light transmittance as well as smoother surfaces. The continuous process provides more uniform thickness and has less of a tendency to form warped sheet.

Continuous cast sheet can be rolled into reels containing 250 to 600 ft of material in length and up to 9 ft in width, and mounted on pallets. Tissue interleaf is used to protect against scratching. The reels are enclosed in a polyethylene over wrap and paperboard package for shipping.

Casting Rod and Tube

Acrylic rods are usually cast in aluminum or nylon open-end cylindrical molds or in tubular bags made from plastic films. The molds are held vertically in a heated water bath. Monomer or syrup is poured into the mold and cured in layers from the bottom. The mold is slowly lowered into the bath at a rate that starts the reaction of a new section before the previous one is completely cured so that boundary lines do not form in the rods. Because of the effects of shrinkage, the cured rods must be finish-ground to round them to true dimensions.

Tubes are made in a spinning metal mold holding the monomer or syrup. The mold rotates in a heated water bath and the centrifugal effect forces the material against the wall where it polymerizes. Wall thickness is controlled by the amount of material, the heating cycle, and the spinning rate. The cured tube is stripped from the mold and polished.

EMBEDMENT

Extensive use is made of embedding (encasing) all kinds of products in acrylic plastic for decorative, safety, or study purposes. They range from jewelry to biological specimens. To obtain a clear bubble-free embedment, certain products such as biological specimens must be dehydrated by careful heating in an oven or by freeze-drying under vacuum. The dried material is then impregnated with catalyzed monomer or casting syrup to fill the space previously occupied by moisture or air. A base of gelled polymer is made by heating syrup in a mold. The specimen is then put on the base, and the mold is filled with polymerizable acrylic. After degassing, the casting is cured by heating in an oven or autoclave.

Thick or large objects can be embedded in acrylic in several stages by curing successive charges of syrup until the needed thickness is attained. Stresses are relieved, particularly in hard castings, by annealing at 225°F for at least 2 hours for each 0.5 in thickness. Soft encasements may be cooled rapidly, but hard castings must be cooled slowly in the mold before stripping to prevent setting up stresses. Cured castings can be machined and polished to heighten their brilliance. Scientific specimens are sliced on a microtome, the slices are mounted on slides, and the plastic is leached away by soaking in acetone (chapter 22).

FILLED CASTING

Filled acrylic castings have been used for marbleized vanity tops, furniture, and other applications. They are made by a relatively simple process and resist burning and the effect of alternate wetting and drying. Syrups for the process are supplied in two components, one pure acrylic and the other acrylic containing a substance that accelerates the decomposition of peroxide catalyst. Each component is mixed separately with usually 1.5 times its weight of alumina hydrate filter; peroxide paste is also added to the pure acrylic.

A method used to produce products takes the two parts that are degassed in a vacuum and combines them in a static mixer to form a reactive mix into which pigment is dripped to provide color and the veined appearance characteristic of marble. Molds pretreated with a release coating of polyvinyl alcohol or a permanent fluoroplastic are filled with the active mix and allowed to cure. The filled castings set in about 30 minutes and subsequently may be post-cured in a forced-draft oven to expel any unconverted monomer. The products are then trimmed and sanded to a smooth matte finish.

Another type of filled casting can be used to repair and protect concrete surfaces. A special acrylic syrup is mixed on site with a blend of inert and reactive fine fillers to a smooth mortar. The mortar is applied to the surface within 15 minutes of preparation. At ambient temperatures of 25°F or higher, the patch or overlay sets hard enough in 1 to 3 hours. The physical and mechanical properties obtained in 24 hours are far better than those of conventional or modified concrete after a 28-day cure. The new surfaces are durable and have been demonstrated to resist the effect of traffic and environmental factors for many years. Note that other plastics are also used in this application (chapter 10).

PROTOTYPE CASTING

Casting of plastics is used to develop prototypes of products. Casting can quickly produce prototype molds, dies, and products from a computer-aided design (CAD) database. It follows systems such as modeling stereolithography, in that it builds the part by layers from a CAD file. The process builds up a prototype one layer of plastic. Details on the process are given in chapter 17.

CASTING OF NYLON

INTRODUCTION

Monomers of the lactam family are used to make cast nylon. They will polymerize under various conditions. Cast nylon continues to offer advantages in applications where wear resistance, resiliency, strength, chemical and abrasion resistance, and lightness are important. Due to the higher molecular weight and crystallinity of cast nylon over that of extruded or injection molded nylon, the cast plastic possesses a higher modulus, a higher heat deflection temperature, improved solvent resistance, lower moisture absorption, and better dimensional stability. In addition, the process of nylon monomer casting is more economical than extrusion or injection molding methods, and both of these latter methods, unlike casting, are restricted to light and thin shapes.

Monomer casting is effective for the fabrication of shaped products in practically all sizes and thicknesses. It also provides economic advantages in low- or high-volume manufacture. Cast parts can be either produced to size or they can be cast and then machined to strict tolerances as required in accordance with end-use needs.

Applications include heavy-duty rotational bearings, gears, sheaves, and slide bearings. These cast nylon parts take advantage of the material's high wear resistance, strength, and lubricity. There are very large bearings of cast nylon used to reduce overall weight by many hundreds of pounds when substituted for bronze and other metal bearings.

The energy-absorbing quality of cast nylon has reduced downtime on a transfer line of a vehicle manufacturer when used for axle pallets. Chain wear strips in caustic soda facilities and pulleys for overhead gates in dilute chlorine environments utilize nylon's chemical resistance. Cast nylon sheaves for cranes extend the service life of wire rope, bearings, and the sheave itself, while increasing lifting capacity through weight savings. Cast nylon gears, rollers, bushings, and bearings are additional applications that run quietly and provide sound absorption. They also reduce the need for oil, grease, and other lubricants.

PROCESS

Casting nylon is a four-step process: melting the monomer, adding the catalyst and activator, mixing the melts, and casting. Optimum melt temperature must be maintained throughout the process. Lactam flakes must be melted to liquid form under controlled temperature and atmospheric humidity. Hygroscopic flake lactam must be protected from excess moisture that would cause the catalyst to decompose, preventing complete polymerization.

Two equally divided solutions of lactam melt are prepared, one containing the catalyst, the other the co-catalyst or activator. The solutions are then thoroughly mixed before being poured into the molds. Anything added to the melt must be thoroughly dried first to prevent incomplete polymerization. Colorants must be selected carefully, because many commercial dyes are not suitable for use with nylon. Some are not soluble in the melt, are not chemical-resistant, are unstable at the operating temperatures involved, or interfere with polymerization. Additives that are not soluble, such as dyes and certain fillers, can present a processing problem because of the difficulty in reaching a homogenous dispersion in the melt.

A scale, electric hotplates, stainless steel pails, thermometers, and stirring rods are the necessary basics for setting up a bench-top operation for prototypes or limited production runs. Melting and adding the catalyst ideally take place in an inert gas atmosphere. The aim is to keep the containers covered as much as possible to reduce moisture absorption in the nylon compound. Bulk handling systems use thermostatic controls, an inert gas system for atmospheric control, and stainless steel vessels with a means for constantly stirring the melt. It is possible for systems to be designed either for gravity or mechanical transfer of the melt into the molds.

Molds must be capable of containing a low-viscosity liquid at temperatures of 200°C. They should allow for shrinkage. Two-piece molds are most commonly used for simple shapes, but more complex molds that are disassembled to remove the cast shape also are constructed. Standard molds for simple shapes are available at very low cost, while complex custom molds can range up to thousands of dollars. Simple molds can be made from silicone rubber, epoxy, or sheet metal, while expensive tool steels generally are used for the more complex molds (chapter 17).

As reported in other sections of this book, annealing plastic may be required. Stresses can develop during the casting process, which could cause brittleness or other difficulties during machining or use of the part. This problem can be minimized, though, by very slow cooling of the cast part. Packing in insulated material for up to 24 hours or cooling in an oven are two methods often used. The best results are obtained by air-annealing in circulating ovens or annealing in an oil bath using highly refined mineral oil.

SOLVENT CASTING OF FILM

In PVC, acrylic, and other plastic solvent casting film the formation depends on solubility, not melting of the plastic. The process thus requires only moderate amounts of heat. The following review concerns using PVC. An example of a solvent used with PVC is tetrahydrofuran (THF). To produce PVC films by this process, the resins, plasticizers, and other materials are added to the solvent in an inert gas-blanketed mixing tank. Thorough mixing with uniform viscosity of the solution, followed by thorough degassing, are critical factors for producing a quality film.

After the solution is mixed and then cooled below its boiling point, it is pumped to the casting tank. The solution is filtered to 5 microns to remove any undissolved particles and is then pumped to a specially designed flat die, where it is cast onto a stainless steel conveyor belt. The belt then enters an oven where the solvent is evaporated from the film. The film is then cooled, stripped from the belt, and wound into rolls.

The die opening, the pumping pressure, and the speed of the belt control the gauge of the film. In-line monitoring equipment is used for gauge control and for quick gauge changes. Heated air traveling counter to the direction of the conveyor belt carries the solvent vapors from the drying oven to the solvent recovery system through large ducts.

Different solvent recovery systems are used. As an example there is the solvent system that consists of fixed bed adsorbers containing activated carbon and a distillation system. The carbon adsorbs the solvent vapors. Then the beds are steamed in sequence to remove the solvent. The solvent and steam are condensed into a large tank. The distillation system is then used to distill the solvent from the water to a purity of 99.99% so that it can be reused. Because of the high cost of solvent, complex monitoring equipment is used to insure a high rate of recovery.

Since high temperatures are not required to dry the film, stabilizers, plasticizers, and lubricants do not have to be added for processing. In addition, any polymer that is soluble in the solvent (usually THF) and that will not adhere to the stainless steel belt can be alloyed with PVC or cast by itself. Typical examples of such polymers include butadiene rubber, acrylics, ethylene vinyl acetate (EVA), and Sarans.

Special PVC resins provide wide and low heat-sealing ranges in rigid films. For example, an unplasticized film can be cast with a heat seal range of 250°F to 340°F or a plasticized type from 180°F to 240°F for use in flexible packaging laminations or for sealing to rigid vinyls.

Films made by solvent casting have sparkle and clarity, good gauge control, low strains, freedom from pinholes, uniform strength in both directions, and good optical properties. Typical applications are flexible packaging laminations for food and drugs, cap stock for scaling to rigid vinyl cups, decals, optically clear storm window film, low-temperature adhesive films, surgical drapes, and unit-dosage liquid medicine cups. Certain cast PVC films also can be processed further by tentering to provide shrink films for food and drug packaging (chapter 5).

While the capital equipment for solvent casting is expensive and the process is considerably more complex than extrusion or calendering, certain types of film can be produced that would be difficult or impossible to manufacture by any of the other film processes.

CHAPTER 12

REACTION INJECTION MOLDING

INTRODUCTION

Developed in 1969, reaction injection molding (RIM) is a relatively new manufacturing technology used to produce high-quality, principally polyurethane (PUR) thermoset (TS) plastic or thermoplastic (TP) parts. Despite its young age, this technology has become a premier plastics molding process that offers versatility in processing options and chemical systems used to produce high-quality, highly styled plastic products. Figures 12.1 and 12.2 provide schematics of the typical RIM process (365, 366).

RIM is a process in which two or more liquid intermediates (isocyanate and a polyol) are metered separately to a mixing head, where they are combined by high-pressure impingement mixing and subsequently flow into a mold, where they polymerize to form a molded part (chapter 1). Although automotive front- and rear-end components compose the largest market, a wide variety of other applications exist. Advantages that are inherent in the process fall into three general categories: low pressure, low temperature, and the use of a reactive liquid intermediate. RIM is also called reactive injection molding. If a plastic system of the RIM type is sprayed against the surface of an open mold, the expression *reactive spray molding* (RSM) is used (13, 212).

With the pressures in the mixing head between 1500 and 3000 psi, the in-mold pressures are significantly lower than in many of the other molding processes. When comparing a typical RIM in-mold pressure of 50 to 150 psi with the 5000 to 30000 psi required for TP injection molding (chapter 4), it becomes apparent why RIM is particularly suitable for larger parts. Automotive bumpers are routinely produced on RIM presses with 100 to 150 tons of clamping force, while comparable injection molded parts require presses of 3500 tons or more.

The temperatures used in RIM are also significantly lower. With PURs, the intermediates normally are processed at temperatures between 75°F and 120°F and the mold is usually between 130°

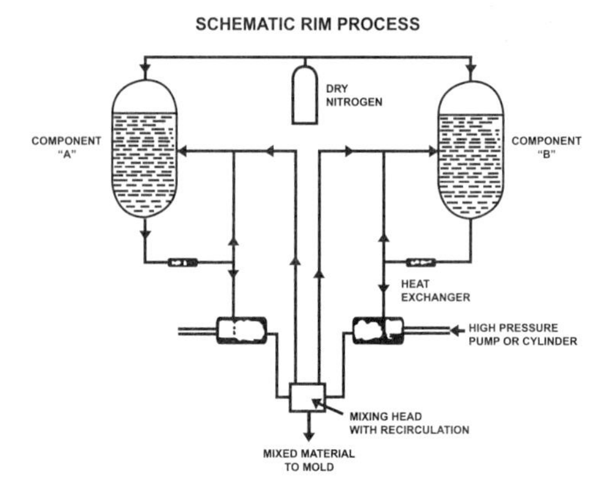

Figure 12.1 Example of typical PUR RIM process (courtesy of Bayer).

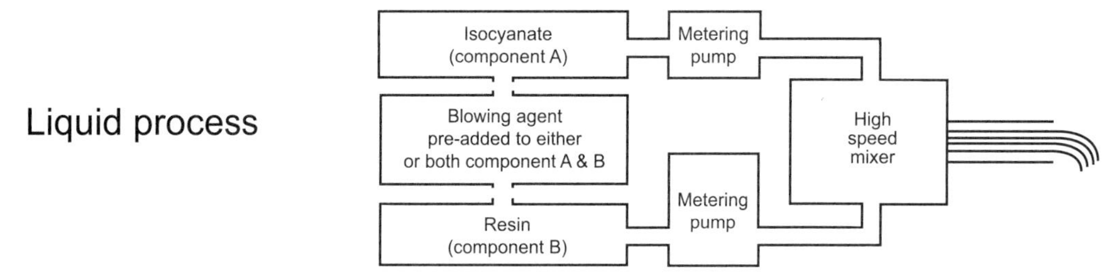

Figure 12.2 Diagram highlighting material use and handling in a PUR RIM process (courtesy of Bayer).

and 170°F. These lower temperatures obviously require significantly less energy consumption than competitive processes.

The use of liquid intermediates has additional benefits beyond the low pressures and temperatures involved. A tremendous amount of design flexibility is possible with RIM. Since the mold is filled with low-viscosity liquid, very complex part configurations can be produced. Ribs, mounting bosses, slots, and cutout areas are all possible. RIM parts can be molded with wall sections as thin as 0.100 in and as thick as 1.5 in (chapter 19). Also, moldings can incorporate variations in thickness within the same part. Incorporation of inserts for mounting or reinforcement is also practical. Since the mold is filled before polymerization occurs, there are no molded-in stresses to cause parts to warp or crack after the demolding process.

The relative ease of compounding TS and TP liquid formulation allows a great deal of flexibility in fine-tuning the material to the requirements of the part. By changing variables such as filter type and level, blowing-agent concentration, pigment, and catalyst, the properties of the plastic can be optimized for the specific application.

The two major classifications for RIM products are (1) high-density, high-modulus, flexible elastomers and (2) low-density structural foams. Automotive trim and fascia are usually elastomers. Furniture and equipment housings are frequently molded as structural foams (especially when texture and/or sound deadening are included in the product specifications).

About 85% of processed PURs are elastomeric. The rest are rigid, usually structural foams that have a solid skin encasing a foamed core. PURs can be used with physical blowing agents such as halocarbons (chapter 8). Foaming is an integral part of the RIM process even for solid products, because it compensates for the shrinkage that occurs during polymerization. That is why most elastomeric products also include foaming agents. The same approach is used during injection molding of solid plastics, in which up to 5 wt% of a blowing agent is used to compensate for shrinkage.

Overall, RIM has advantages over the standard low-pressure mechanical mixing systems in that larger parts are possible, mold cycles are shorter, there is no need for mold solvent-cleaning cycles, surface finishes are improved, and rapid injection into the mold is possible. Large and thick parts can be molded using fast cycles with relatively low-cost materials. If surface coating is required, the types of coatings used are coating paint, in-mold coating (Fig. 12.3), film, and metallic facings. RIM's low energy requirements and relatively low investment costs make it attractive. Applications are many; they include automobile bumpers, medical products, radio and TV cabinets, furniture, sporting equipment, appliances, and housings for business machines.

An example of a medical product is shown in Figure 12.4. This PUR RIM product from Thieme Corp., St. Charles, Illinois, is possibly one of the largest known single-shot RIM parts. It is a twelve-part enclosure for a computerized tomography (CT) device. When all twelve RIM parts are assembled, the enclosure is large, as described in Table 12.1. Many of these parts use a reinforcing rib design by Thieme that provides support and rigidity and allows the assembled CT unit to be moved.

Figure 12.5 shows an appliance application for RIM. Italian molder GMP Polyurethanes SpA created a refrigerator door that is as much a fashion statement as it is functional. GMP developed a new surface-finishing technique that takes advantage of the outstanding adhesion between PUR and film. The patented process cuts costs by eliminating the need for postpainting, while at the same time achieving an improved surface finish.

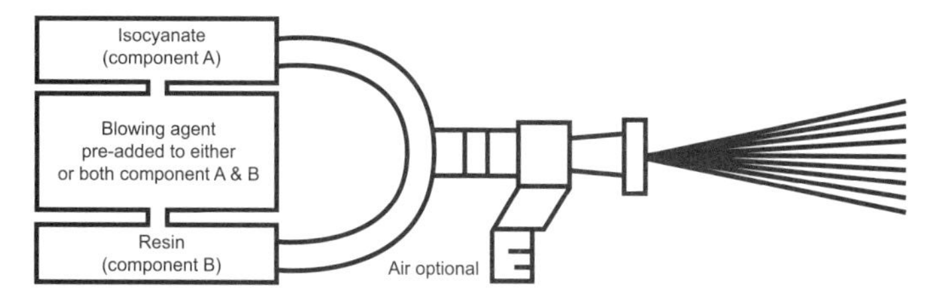

Figure 12.3 Example of in-mold coating application.

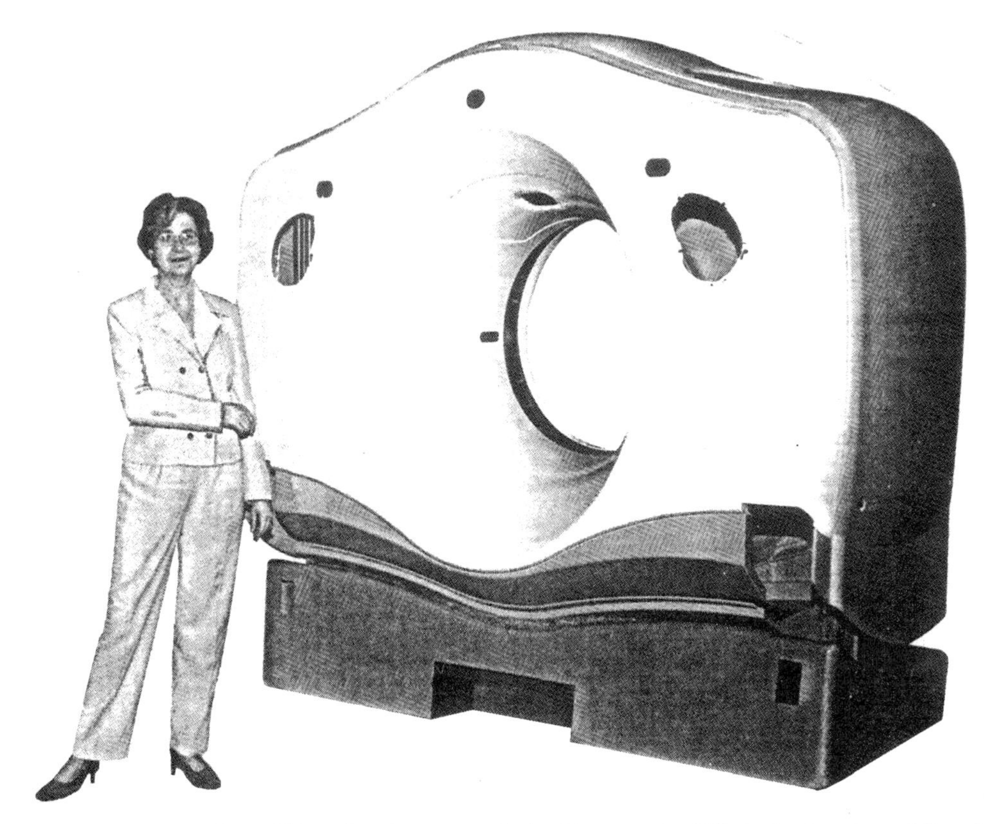

Figure 12.4 Polyurethane RIM product for a computerized tomography (CT) device (courtesy of Bayer).

Material: Baydur® 110 polyurethane structural foam RIM system

Molder: Thieme Corporation

Weight: 74 pounds (rear panel)

Dimensions: 94 inches wide, 74 inches tall and 39 inches deep (when assembled)

Key properties: Strength, excellent surface finish and large-part capability

Innovations: Believed to be one of the largest-known single-shot RIM parts

Table 12.1 Information on computerized tomography (CT) devices (courtesy of Bayer)

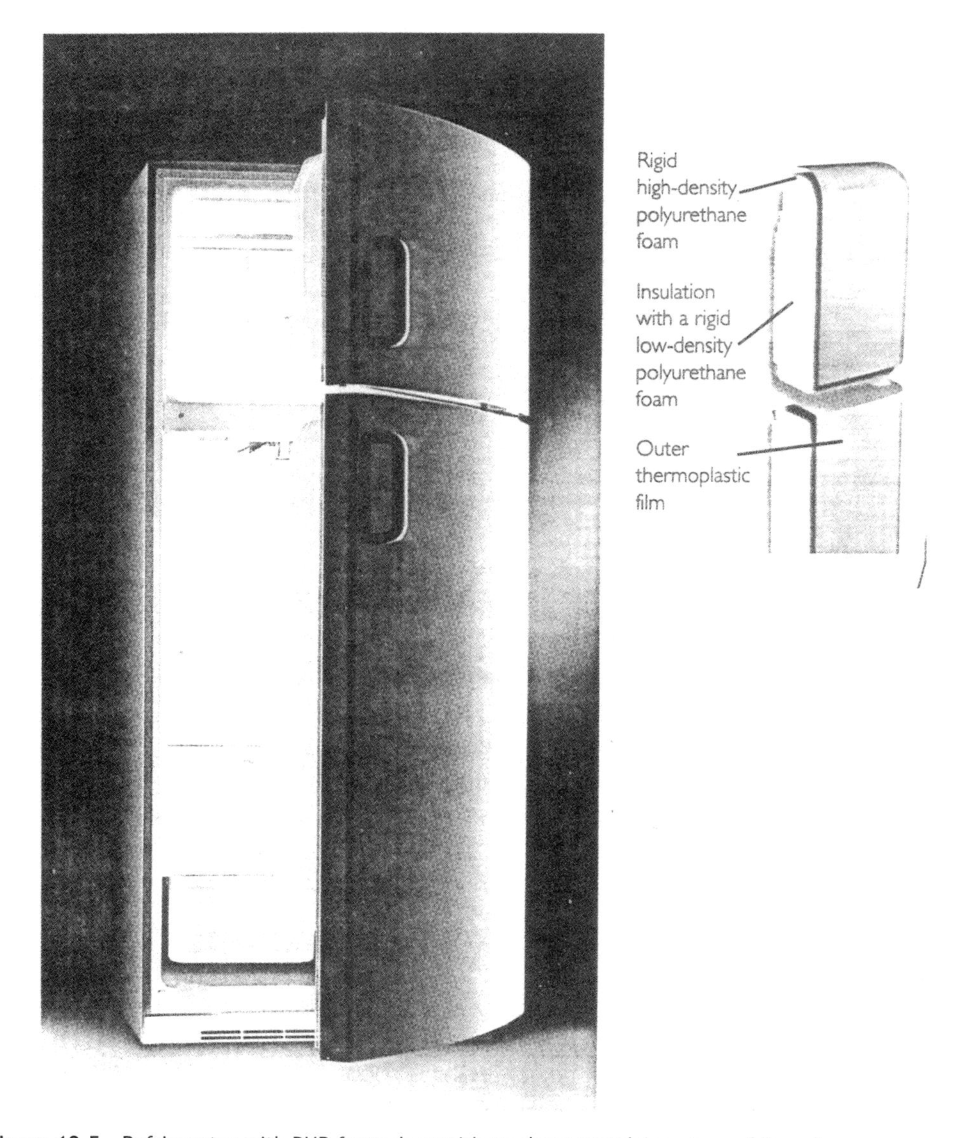

Figure 12.5 Refrigerator with PUR foam door with no sheet metal (courtesy of Bayer).

The GMP's process eliminates the use of sheet metal for the skin of the refrigerator door (Table 12.2). In this application, the TP film forms a durable, protective outer skin with a wide choice of color options that are applied directly to the film. In addition, more innovations exist apart from the film and TP interior liner. The doors consist entirely of PUR. GMP backs the TP

- Cut thermoplastic film to size and place it in the mold
- Thermoform the film in the mold until it is an exact image of the surface of the mold
- Inject polyurethane RIM system into the same mold to create the substrate
- Demold the part
- Deflash the part

The Benefits of this are obvious:

- Shorter production time due to elimination of post-painting
- Cost reduction leading to greater competitiveness
- Greater freedom of design than with conventional sheet metal skin

Table 12.2 Information on GMP's patented refrigerator door technique

film with an approximately 4-mm-thick layer of the Baydur 110 structural foam PUR RIM system from Bayer AG, which creates a rigid, dimensionally stable outer shell with no need for sheet metal. Then, GMP fills the space between this shell and the inner liner with insulating PUR that is rigid, low-density foam. The result is a self-supporting door that satisfies all stability, thermal insulation, and surface finish requirements.

EQUIPMENT

Processing equipment consists of the material-conditioning system, the high-pressure metering system, the mixing head, and the mold carrier. Since the RIM process involves a chemical reaction in the mold after the intermediates have been mixed, for consistent parts to be produced, the material delivered to the mix head must be consistent from shot to shot.

The material-conditioning system is designed to ensure that the materials fed to the metering pumps meet these requirements. It typically includes tanks to hold the intermediates, agitators to ensure that the material in the tanks is of homogeneous temperature, and a nucleation-control system that keeps the level of dissolved gases in the polyol component at the desired level. The tanks can range in size from 15 to 150 gal or larger depending on the consumption rate. These tanks are normally automatically refilled at frequent intervals from bulk storage tanks. Jackets on the tanks as well as heat exchangers on circulating loops are used for temperature control (Fig. 12.2).

The metering system takes the conditioned intermediates from the supply tanks and delivers them to the mixing head at the desired rate and pressure. There are two basic types of metering systems: high-pressure axial or radial piston pumps, and lance displacement cylinders. The piston pumps are hydraulic pumps that have been modified to handle chemicals. They are capable of continuously metering at pressures up to 3500 psi. Lance pistons, which are driven by a separate hydraulic pump, displace the reactants from a high-pressure metering cylinder. In addition to more precise metering, they have the capability of processing filled systems.

The mixing head contains a cylindrical mixing chamber, where the intermediates are mixed by direct impingement at pressures ranging from 1500 to 3500 psi. It also contains a cylindrical cleaning piston that, after the shot is complete, moves forward to wipe the remaining materials out of the mixing chamber (otherwise the mixing head would have cured plastic, preventing the mixing of the next shot). There is a valving mechanism to shift the material flow between recirculation back to the tank and flow into the mixing chamber. This action allows the circulating materials to reach equilibrium at the proper temperature, pressure, and flow rate before shifting into the mixing position.

The mold carrier holds the tool in the proper orientation for molding, provides enough clamping force to overcome the in-mold pressure, opens and closes the mold, and positions the open mold in an accessible position for demolding, cleaning, and preparing the mold for the next shot. There is a wide variety of designs and sizes available. Examples of RIM mold–carrying equipment in use are reviewed in Figure 12.6 through Figure 12.10.

MOLD

Since the in-mold pressures in RIM are generally relatively low (50 to 150 psi), a variety of tooling constructions have been used. These include machined steel or aluminum, cast aluminum or Kirksite, sprayed metal or electroplated shells, and reinforced or aluminum-filled epoxy (chapter 17). With mold pressures usually below 100 psi, mold-clamp-pressure requirements can accordingly be low when compared to those used in injection and compression molding.

Some of these constructions are relatively inexpensive when compared with other large-volume production tooling. The low-viscosity liquid that fills the mold, which is heated to 120°F to 160°F (49°C to 71°C), will duplicate exactly the surface of the tool. Consequently, when good surface characteristics and high tolerances are required, machined tooling has generally been the chosen route, particularly for higher-volume production runs. The ability to use less costly tooling methods for prototype and for short runs, however, remains a significant advantage of the RIM process.

Since one of the ultimate objectives of the RIM process, for its major market of automotive exterior part production, was a cycle time of 2 minutes or less, a great deal of effort was applied to mold construction and design. Continuous automatic operation of a molding station without interruption required improvements in mold release and mold surface technology. Originally, mold preparation following a shot was required due to the buildup of external release agents, which were necessary to enable easy removal of the part from the mold.

Figure 12.6 RIM machine with mold in the open position (courtesy of Milacron).

This problem was approached from the material side, through a search for suitable internal releases, and through the development of improved external mold release compounds. From the equipment side, the development of automatic molds was required if the RIM process was to compete with classical injection molding with respect to mold cycle times and efficient production.

General Motors Corporation constructed such a mold for a production trial of the 1974 Corvette fascia (which actually started the development of RIM). This mold was tool steel with a highly polished nickel-plated surface. Most of the mold seals were elastomeric, to prevent excessive flash (up to 10%, by weight, of flash can occur; PUR could not be reused, since a TS was used) due to leakage of the low-viscosity TS PUR reacting material. This was possible because the internal mold pressures encountered in the RIM process were less than 100 psi. This evaluation was highly successful in demonstrating the capability of total automation of the RIM process.

Figure 12.7 RIM machine with mold in the closed position (courtesy of Milacron).

Figure 12.8 Example of an auto bumper RIM production line (courtesy of Milacron).

Figure 12.9 RIM machine with auxiliary clamping system (courtesy of Battenfeld).

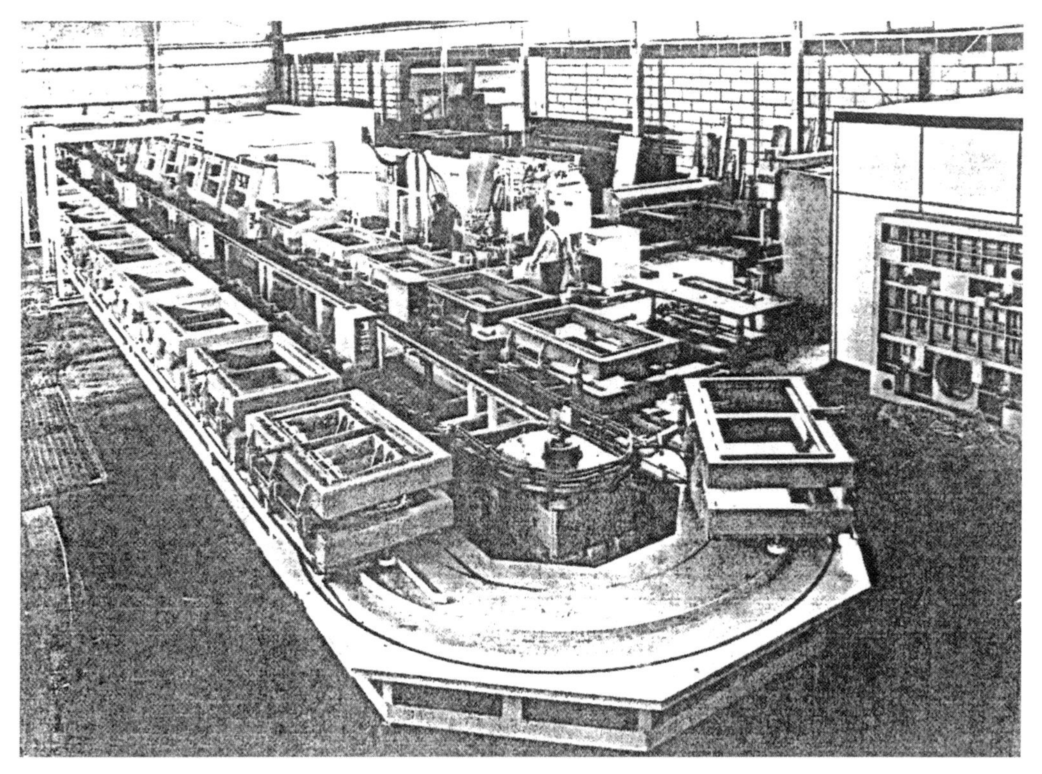

Figure 12.10 Example of a RIM production line, where molds are on a moving track permitting final cure of PUR (courtesy of Battenfeld).

In the construction of molds for RIM processing, parts quality and finish are roughly equivalent to the quality and finish of the mold surface itself. A common misconception is that because the clamp tonnage for an RIM setup is relatively low, low-quality tools can be used. This, however, is true only insofar as the pressure requirements for the mold are concerned.

Experience has shown that the finish on the part surface is a direct function of the mold finish and that the mold finish is a direct function of the quality of the mold material. Excellent results have been obtained using high-quality, nickel-plated, tool steel molds and electroformed nickel shells.

For production runs of 50,000 parts per year, a P-20, P-21, or H-13 steel would be most appropriate, not only because of these steels' homogeneous nature, but also because they are able to be easily polished and are adaptable for a good planting job. The prehardened grades of 30 to 44 Rc are preferable because of the degree of permanency that they impart to a tool. After machining, a stress-relieving operation is very important in order to avoid possible distortions or even cracking (chapter 17).

Nickel shells that are electroformed or vaporformed when suitably backed up and mounted in a frame are also excellent materials for large-volume runs. For activities of less than 50000 parts per year, aluminum forgings of Alcoa grade 7075-T73 that have been machined to the needed configuration will perform satisfactorily. They have the advantage of good heat conductivity, an important feature in RIM.

Cast materials are used for RIM molds with reasonable success. One such material is Kirksite, a zinc-alloy casting material. Kirksite molds are easily casted, are free from porosity, will polish and plate well, and have been used with favorable results.

The mold temperature should be maintained within ±4°F for consistent quality and molding cycles with PUR. The mold temperatures will range from 100°F to 150°F, depending on the composition being used. The cooling lines should be so placed with respect to the cavity that there is a ¾ in wall from the edge of the hole to the cavity face. The spacing between passages should be 2.5 to 3 diameters of the cooling passage opening. These dimensions apply to steel; for materials with better heat conductivity, the spacing is usually increased by one hole size.

As with the chemical components, it is necessary to maintain constant surface temperatures in the mold for a reproducible surface finish and constant chemical reactivity. This temperature varies according to the chemical system being used and has been determined empirically.

The mold orientation should allow filling from the bottom of the mold cavity, so that air can escape through a top flange on a hidden surface. This allows controlled venting and the positioning of vent pockets that can be trimmed from the part at a later time.

RUNNER AND GATE DESIGN

Figure 12.11 through Figure 12.17, along with Table 12.3 and Table 12.4, provide an introduction to designing the RIM melt flow from the mixer into the mold cavity (366).

COST

Low-cost tooling is a primary benefit of RIM, especially for start-up companies and those needing only a small quantity of parts for a particular product line. Tooling costs are lower because machine pressure is much lower compared to high-pressure molding at several thousand psi, and because

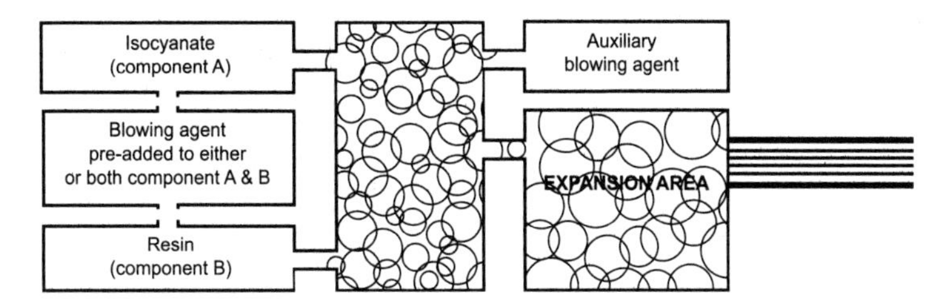

Figure 12.11 Gating and runner systems demonstrating laminar melt flow and uniform flow front (courtesy of Bayer).

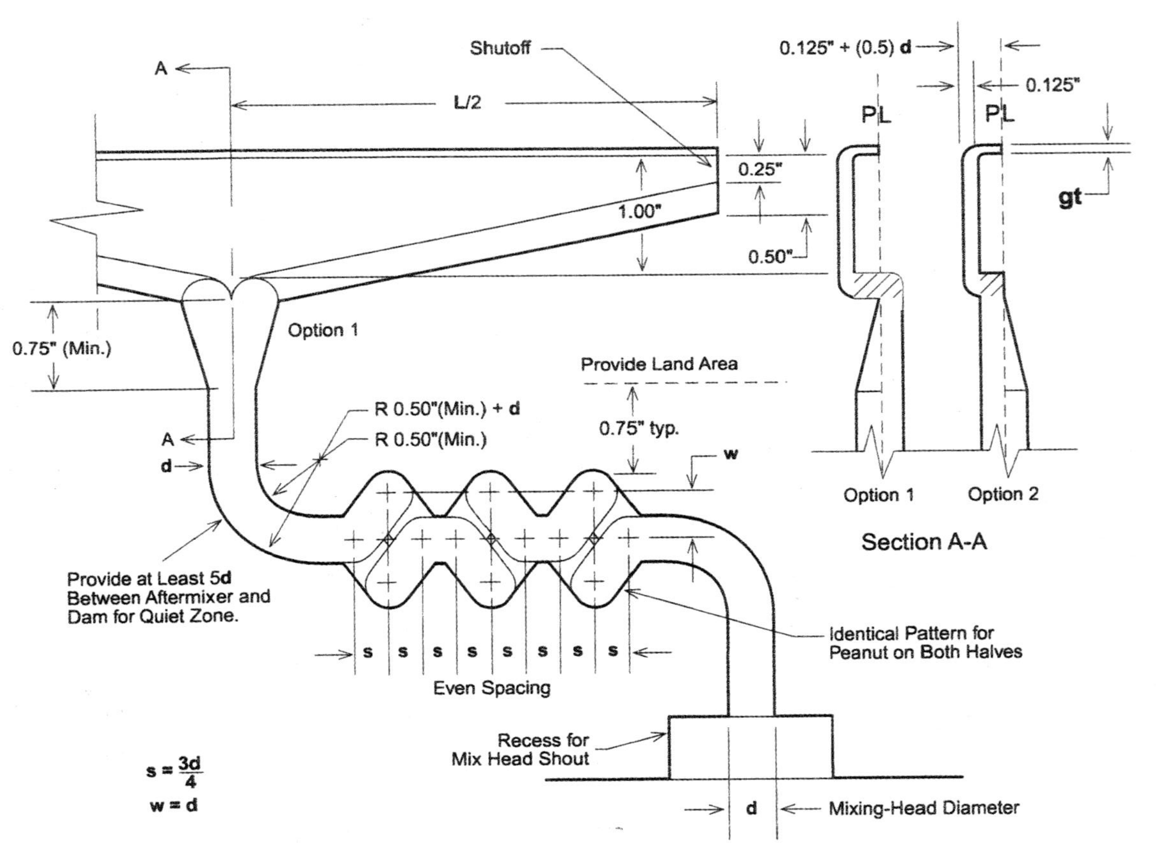

Figure 12.12 Example of a dam gate and runner system (courtesy of Bayer).

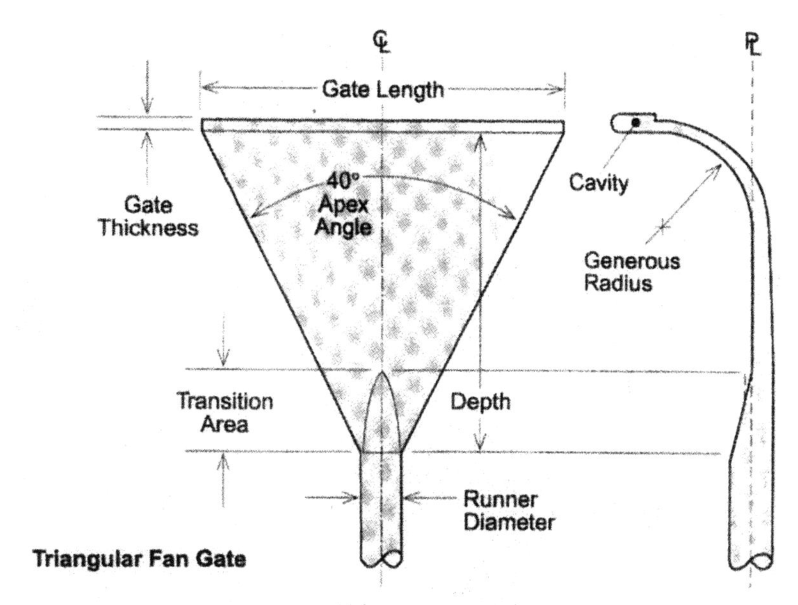

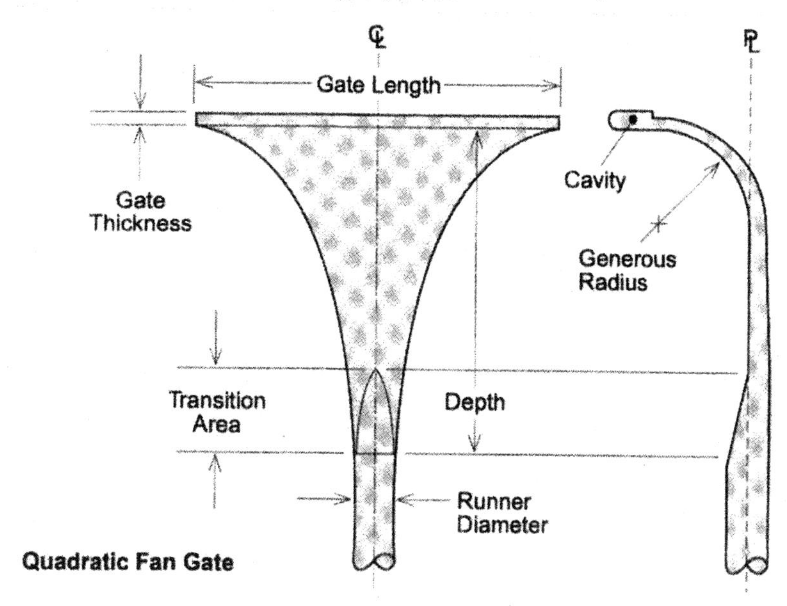

Figure 12.13 Examples of triangular and quadratic fan gates (chapter 17; courtesy of Bayer).

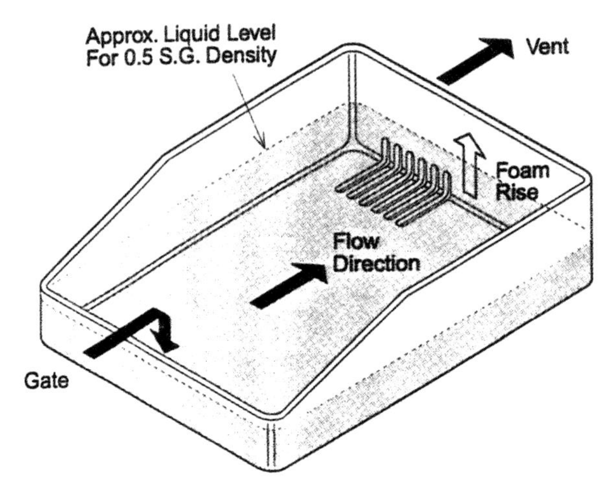

Figure 12.14 Example of melt flow around obstructions near the vent (courtesy of Bayer).

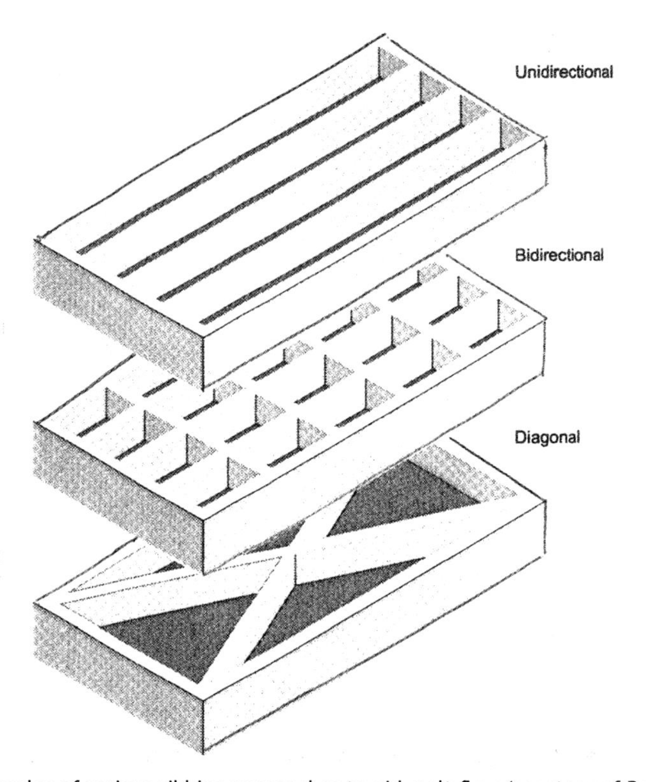

Figure 12.15 Examples of various ribbing approaches to aid melt flow (courtesy of Bayer).

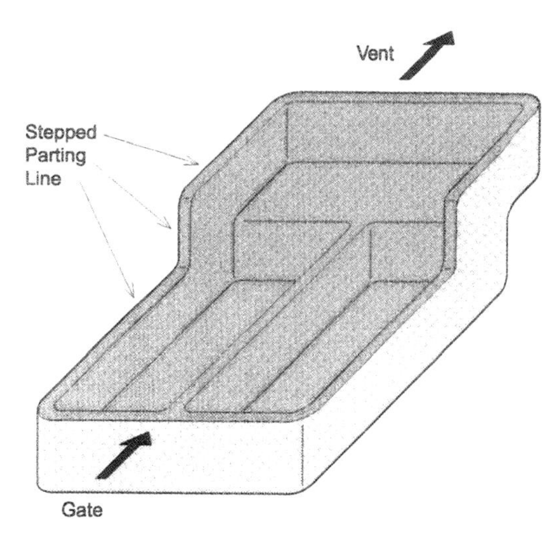

Figure 12.16 Example of a low gate position with high vent for best results when foaming (courtesy of Bayer).

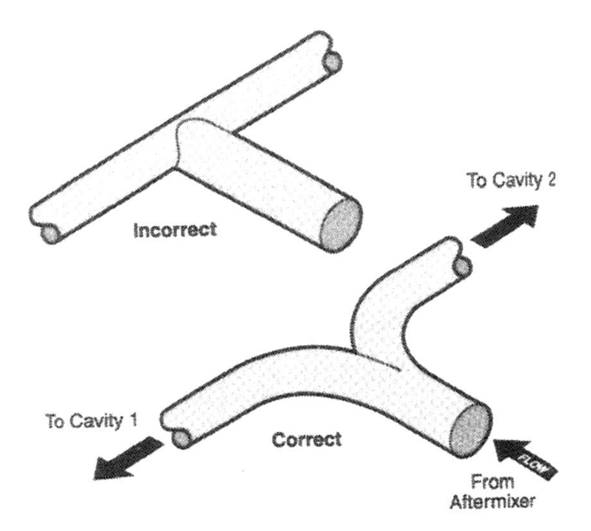

Figure 12.17 Example of how to properly split a melt stream from the mixer (courtesy of Bayer).

molds are only heated to 170°F instead of 200°F or higher. Molds can be made from several different materials that offer different price ranges (561).

The RIM process allows fabrication of small and large parts with equal ease. Large parts include those that are at least 6 ft long, 4 ft wide, and 4 ft tall, with a special holding clamp of 5 ft by 5 ft in size that is used on the molding machine. The process lends itself to reduced setup time and

Known Variables	Solid Systems	Foamed Systems
Weight Output: O_W	lb/sec	lb/sec
Material Density: D	lb/ft^3	lb/ft^3
Maximum Injection Velocity: V	15 ft/sec	5 ft/sec
Gate Thickness: gt	0.10 in	0.08 in
Calculated Variables		
Volumetric Output: $O_V = O_W / D$	ft^3/sec	ft^3/sec
Minimum Cross sectional Area:		
$A = (O_V \cdot 144) / V$	in^2	in^2
Minimum Gate Length: $L = A / gt$	in	in

Table 12.3 Calculations for determining dimensions for a dam gate (courtesy of Bayer)

costs, and consequently makes just-in-time delivery easier. Machine tooling can be made out of hard epoxy, which is still very accurate tooling since it comes from computer-aided design (CAD) models (chapter 25), or different grades of aluminum are used for making molds. Use is made of the softer materials that still produce precision molds because of the low heat and low pressure required.

A disadvantage of the RIM process is that per-part costs are higher. However, on small runs of a few hundred parts, the lower cost of tooling far outweighs the cost of parts. If a customer requires several thousand parts per month, it is less costly per part to use expensive, hard steel tooling with high-pressure molding. Another downside to the RIM process is that it is limited to using a few plastics with principally PUR, whereas many different plastic materials can be used with high-pressure molding. However, PURs are available with different material properties.

PROCESSING

This high-pressure impingement mixing, which delivers two or more liquid urethane components to a very small mixing chamber that continuously mixes and injects into a closed mold, delivers at rates approaching 650 lb/min. The liquid components are heated to maintain low viscosities.

The heart of the system is the mixing chamber, where the liquid components must be thoroughly mixed without imparting turbulence. High-volume, high-pressure recirculating pumps from liquid-storage tanks accomplish continuous delivery of the components to the mold. Automatic controls are used to maintain precise flow and temperature of the plastic.

Unlike injection molding, RIM does not require the clamping press to be close to the material source. The components can be transferred safely across the floor of the processing plant. A

Known Variables	Solid Systems	Foamed Systems
Weight Output: O_W	lb/sec	lb/sec
Material Density: **D**	lb/ft^3	lb/ft^3
Maximum Injection Velocity: **V**	25 ft/sec	5 ft/sec
Gate Thickness: **gt**	(specify) Typically gate thickness depends on the part thickness at the location of the gate for a solid polyurethane elastomer or solid polyurethane structural material. For polyurethane structural foam, the typical gate thickness does not exceed 0.080 inches.	
Calculated Variables		
Volumetric Output: $O_V = O_W / D$	ft^3/sec	ft^3/sec
Final Fan Area: $A_F = (O_V / V) \cdot 144$	in^2	in^2
Gate Length: $gl = A_F / gl$	in	in
Initial Fan Area: $A_O = pd^2 / 4$	in^2	in^2
Original Fan Thickness: $t_O = d / 2$	in	in
Original Fan Length: $L_O = A_O / t_O$	in	in
Mid Fan Area: $A_M = (A_F + A_O) / 2$	in^2	in^2
Mid Fan Thickness: $t_M = (t_O + gt) / 2$	in	in
Mid Fan Length: $L_M = A_M / t_M$	in	in
Fan Depth: $D_A = gl - (2pr / 4) + r$	in	in

Table 12.4 Calculations for determining dimensions for a quadratic gate (courtesy of Bayer)

metering unit can accommodate as many as five mixheads or molding stations because the lapsed time for the metering shot is only a small fraction of the overall molding cycle.

The typical PUR RIM process involves precise metering of two liquid components under high pressure from holding vessels into the static impingement mixhead (Fig. 12.18). The coreactants are homogenized in the mixing chamber and injected into a closed mold, to which the mixhead is attached. The heat of reaction of the liquid components vaporizes the blowing agent, beginning the foaming action that completes the filling of the mold cavity.

Proper temperature control of raw material is critical for maintaining the best product characteristics. The ambient temperature in the plant during every season is important to consider when choosing the right heating and chilling equipment to maintain accurate process control of the material. It is important to know the average temperature in the plant year-round to properly size the conditioning system to the process. Most temperature-conditioning systems will require a source

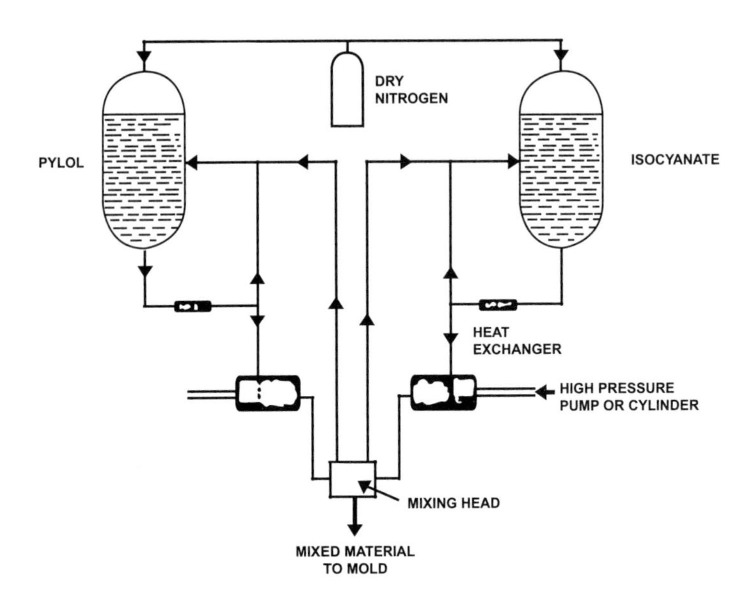

Figure 12.18 Basic schematic for mixing two liquid components to produce a PUR.

of city or chilled water to operate the conditioning equipment. One must make sure this source is available to the metering equipment if required.

Two streams of PUR chemicals collide with each other violently and under high pressure, which is generally at 1500 to 3000 psi inside the mixer. When these impinging streams collide, the flow is very turbulent and the reaction begins. The stream exits the mixhead and is directed into the mold. After the pour a piston inside the mixhead scrapes the walls of the chambers completely clean so that no reacted foam is left inside the mixhead.

There is the straight-through mixhead with its straight chamber into the mold. It has been largely replaced by an L-shaped mixhead with its bent chamber. Processors usually prefer the L-shaped because there is laminar flow when the mix exits the head, and an aftermixing action can be built into the mixing head instead of into the mold (where it occurs for straight-through mixheads).

If the temperature is not properly controlled during mixing, the viscosity of the mix will change, reducing throughput, lowering efficiency, and impairing the quality of and perhaps even damaging the products. A metering unit measures the chemicals and delivers the required amount to the mixhead. Electronics and closed-loop controllers are used for pump-type metering units. Although there are lower-cost systems that can process quality PUR foam, they may not be able to upgrade as requirements increase. For example, the use of smaller tanks may limit the shot-size capability. Throughput for RIM can range from 0.25 to 30 lb/s.

The chemical system and the final molded product requirements determine machinery requirements. Features to review in specifying equipment based on requirements to produce products include the addition of a third- and fourth-component coloring paste in order to mold colored products. Many machinery suppliers offer color-dosing units in conjunction with a three- or four-component mixing head as auxiliary equipment. Clamps come in a variety of shapes and sizes; most are custom-built. A clamp should have a smooth action throughout its entire operational sequence. Any error in movements can damage a mold, and improper sequencing can lead to poor production quality.

Processing of RIM elastomers, as well as structural foams, often includes the dispersion of insoluble gas (air or nitrogen) in the form of small bubbles into the polyol component. This action results in nucleation. It is used to improve the flowability of chemical and to improve the cell structure of the final product. Improved flowability makes the melt flow more laminar and increases the throughput via a very fine pattern of cells all through the molded product.

Molders utilizing this system require equipment to measure and control the amount of entrained gas in the liquid at the desired level. They can include mass-flow meters with density devices, nuclear density-monitoring devices, or a variety of other density-measuring devices to control nucleation level. All these systems work within very defined pressure and temperature limits; however, outside these limits, readings become erratic. There are systems that remove the dependence on system pressure and temperature, providing more consistent data.

With relatively long cure times that are much longer than the duration of the molding cycle, shuttles, turntables, or mold movement tracks (Fig. 12.10) are used as production solutions. Using this type of action in moving the mold, the metering unit can be used to greatest effect, optimizing the time interval between shots. Software programs are available so users can monitor and control the complete process. The software generates a graphic illustration of process parameters such as pressures, temperatures, mixture levels, mixture ratios, and output rates. Software is also available for preventive maintenance and troubleshooting.

PROCESS CONTROL

The chemical systems for RIM all have one characteristic in common: they require an RIM machine to convert liquid raw materials into quality plastic products. Assuming a properly formulated chemical system, the quality of the end product results from the ability to measure, control, and adjust temperature, ratio, pressure, and other essential process parameters of the RIM dispensing machine. Such exacting control leads to a reduction in start-up time, minimal rejects and touch-up work, reproducible product quality, and the ability to pinpoint changes in product properties.

For example, in the high-temperature RIM processing of nylon, temperatures are monitored and controlled within $\pm 2°F$ using both electrical heat tracing and hot oil jacketing. The controllers contain high/low set points; all temperature zones must be at the required settings to permit proper machine operation. A graphic diagnostic panel with light-emitting diodes (LED) associated with all key switches, valves, and pressures, aids in troubleshooting; if a malfunction occurs, a blinking light pinpoints the cause. Low- and high-pressure circulation is monitored by transducers and displayed digitally; high/low pressure limits, if exceeded, will abort the RIM cycle for safety reasons.

One must understand the RIM material chemical system that is directly related to meeting the molded product requirements. One must also take into consideration chemical system properties and terms such as those in Table 12.5 that include nonchemical properties that are also important to understand:

Material properties combined with end-product requirements such as product size, flowability through the mold, and cycle times determine the pressure and output requirements of the processing equipment. To begin selection of the chemical system, one writes a performance specification for the product. Recommended formulations for specific product types have been thoroughly tested and evaluated by the chemical companies that sell them. These companies can provide physical property data for the formulation.

All data must be carefully evaluated in order to make an educated decision about what chemical system to purchase for the production of a specific product. Rough guidelines can be established by knowing what other types of products are manufactured using the chemical system that is being evaluated. The data should be compared to the expected results. Upon completion of a careful evaluation and selection of a chemical system, the next step is to match the process control system with the processing machinery.

MATERIAL

RIM was developed as a processing technique for PUR, and to date the bulk of the usage has been with that material. Fortunately, PURs and related plastics are a tremendously diverse group of materials with a range of properties to fill the needs of very different applications.

These PURs are produced by a volatile chemical reaction. Compounds containing active hydrogens, alcohols in the form of polyols, react with isocynanates in an exothermic reaction to form PUR. This process produces the plastic by starting with the monomer (chapter 1).

The basic materials used are polyols and isocyanates. Polyols may be polyethers or polyesters. The isocyanates may be diphenylmethane-4,4-diisocyanate (MDI) or toluene diisocyanate (TDI). Additives such as catalysts, surfactants, and/or blowing agents are also incorporated. Their purpose is to develop the chemical reaction and form a finished product possessing the desired properties.

The high degree of reactivity of the isocyanate (NCO) group is the key to PUR chemistry. A urethane group is obtained by reacting the isocyanate group with an alcohol (OH) group. To obtain the PUR products discovered by Otto Bayer in 1937, isocyanates with two or more NCO groups must essentially be converted using compounds that likewise contain at least two OH groups (polyols).

All industrial PUR chemistry is based on only a few types of basic isocyanates. The most significant aromatic diisocyanates are TDI and MDI. TDI is derived from toluene. This is initially nitrated to dinitrotoluene, then hydrogenated to diamine, and finally phosgenated to diisocyanate. A defined mixture of isomers comprising toluene-2,4- and 2,6-diisocyanate is obtained (Fig. 12.19). Approximately 1.3 million tons/year of TDI are produced worldwide, most of which are used in the production of PUR flexible foam materials.

1. Viscosity: It is the property of a chemical system that causes it to resist flowing in relation to it's thickness usually measured in centipoise (water has a viscosity of 1 cps).
2. Exotherm The heat which is released as a result of a chemical reaction.
3. Polyol: Chemical compound with more than one reactive hydroxyl group.
4. Impingement: Mixing components using high velocity turbulent contact of two or more component streams.
5. Cream time: It is the measurement of time after material has been dispensed and mixed and the start of its volumetric expansion.
6. Gel time: It is the time required for a reaction mixture to form a gel under specific temperature conditions.
7. Free rise: The unrestricted rise of a foam sample.
8. Peak rise point: The highest rise a foam may achieve prior to the occurrence of cell rupturing and therefore settling.
9. Cure time: The time required for the complete chemical reaction is 100% complete and all desired physical properties are achieved.
10. Tack free time: It is the time between mixing and dispensing, at which the surface of the foam can be touched without sticking.
11. Clean out piston: The hydraulically operated piston which cleans the let down chamber of a mixing head.
12. Cycle: The time for all operations in a manufacturing process. Usually designated as the time span between a step in a molding operation and the subsequent repetition of that step.
13. Demold time: The time from which the reactive materials have been mixed and injected into the mold to a point at which sufficient cure is achieved. This enables removal from the mold without damage.
14. Green strength The strength of a product at the time of demolding. It should be sufficient to prevent any permanent deformation of the product.
15. Post cure: The period of curing after demolding.
16. Back pressure The resulting higher pressure in a line upstream of a restrictive device (mixing head jet).
17. Lead-lag: The problem of off-ratio or non-mixed material that occurs at the opening and closing of a RIM mixing head.
18. Blowing chloride agent: A material normally water, fluorocarbon or methylene that during the reaction either reacts to form carbon dioxide in the case of water or vaporizes from the exotherm that in turn produces the foam.
19. Blanket: A layer of inert (un-reactive) material, normally nitrogen or dry air, placed in tanks at pressure. They are used to avoid contamination of the chemicals from the atmosphere.
20. Day tank: The tanks containing the chemical components usually mounted on or near the machine. Tanks maintain the components at process temperature and feed the machine for production.
21. Degassing Applying vacuum to the day tank in order to remove dissolved gasses and entrapped gas bubbles.
22. Agitator: A rotating or reciprocating device that induces motion in fluid mixtures in order to accomplish uniform dispersing of the components.
23. Adjustable speed: The ability of a machine to increase or decrease the speed of the main drive mechanism. This may be accomplished by methods such as variable pulleys, DC motors, frequency drives and hydraulic or pneumatic flow rates.
24. Adjustable output: The machine's ability to change the flow rate of delivery to the mixing head.
25. Ambient: The room temperature or pressure where the work conducted at a specific time.
26. Cell: Refers to the cavities left in the foam structure after the bubble walls have completely polymerized and solidified. Open cell: Cells interconnect, the structure is rib like (skeletal). Closed cell: Cells are totally enclosed by a membrane and rib structure.
27. Closed mold: A mold in which a mixing head is attached, injects the mixed ingredients into the mold which is fills completely with only gas and minimal flash leakage.
28. Fixed output pump: A pump with an output that is unchangeable.
29. Fixed ratio: Normally refers to a machine that meters materials at only one specific ratio.

Table 12.5 Terminology of chemical and other terms

30. Gate: The opening in a mold leading from the mixing head to the mold cavity. Waste material removed from the gate when the mold is opened may also be referred to as the gate.
31. Open molding: The practice of pouring into a mold that has a removable lid. The lid is closed after the liquids have been poured into the cavity.
32. Positive displacement: Refer-ring to pumps that displace material at a constant rate, over a wide range of conditions with no internal losses.
33. Shot timer: An electronic timer used to control shot size.
34. Temperature stratification: A condition encountered in day tanks when they do not have sufficient agitation to maintain a homogenous state.
35. Throughput: Synonym for flow rate.

Table 12.5 Terminology of chemical and other terms *(continued)*

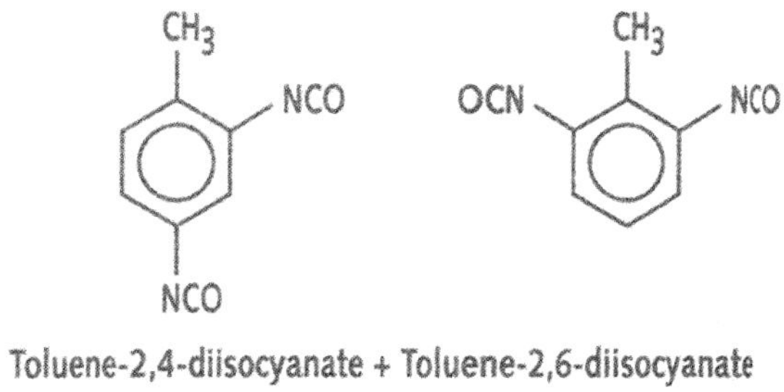

Figure 12.19 TDI is an isomer comprising toluene-2,4- and 2,6-diisocyanate.

TDI production has long since been overtaken by that of MDI, which is currently running at about 2.3 million tons/year (Fig. 12.20). The abbreviation MDI is derived from its former name, methylene diphenyl diisocyanate. MDI is produced by phosgenation of methylene diamine, which is obtained from benzene using nitrobenzene and aniline and by condensing using formaldehyde. The actual MDI is then distilled from the raw phosgenated product, which is extensively present as a mixture of isomers and homologues. It is primarily used for PUR elastomers. The main quantity remains a mixture of compounds with two or more aromatic rings and is known as polymeric MDI (PMDI).

Polyurethane rigid structural foam was one of the earliest applications for RIM. Light weight and rigidity characterize the material. It consists of a solid skin and a lower-density cellular core. Use includes equipment housings, furniture, building components, fancy tires (562), and a variety of industrial and consumer applications (Table 12.6).

Low-modulus elastomers are materials that have found wide use in the automobile industry for fascia, bumper covers, and trim parts. Other applications include integral window seals and replacements for molded rubber. Most of these materials are not pure PUR, but PUR-polyurea hybrids with improved processing and properties compared to those of the earlier all-urethane systems.

There are also high-modulus elastomers. Modifications to the chemistry of producing the low-modulus elastomers allow for the processing of tough polymers with a flexural modulus as high as 250,000 psi. These are used in a variety of large industrial and consumer parts.

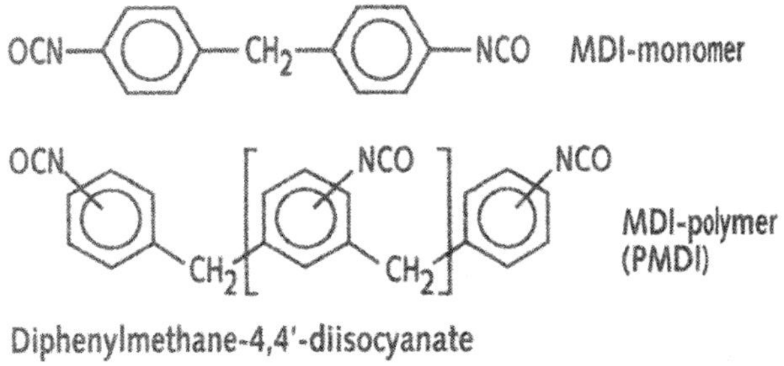

Figure 12.20 Diphenylmethane-4,4-diisocyanate (MDI).

Integral skin foams are useful. They are flexible urethane foams with a high-density skin. They are used in applications that must combine a tough surface and a soft feel, such as steering wheels, arm rests, and protective covers.

One of the leading materials for automotive body panels is the all-polyurea systems that have improved high temperature stability. A very fast reactivity and rapid cycle times characterize these systems.

Heavy trucks and farm tractors contain RIM parts that use poly-dicyclopentadiene (P-DCPD). This material went from industrial applications to heavy-vehicle exterior components competing with fiberglass-reinforced polyester (FRP; chapter 15) and aluminum. The big breakthrough in the heavy truck arena came in 1996 with the Kenworth T2000 eighteen wheeler, which had fourteen exterior components of P-DCPD varying in size from an 80 lb roof fairing to smaller parts of 10 to 15 lb each. In the past couple of years P-DCPD RIM made its mark in the hood of Class 8 heavy trucks.

The first P-DCPD hood appeared in 2001 on the model 9900 from Navistar's International Truck and Engine Corp. in Warrenville, Illinois. Currently the most prestigious application for P-DCPD is the hood on the new top-of-the-line W900L model from Kenworth Truck Company. It replaced a spray-up FRP at no additional cost and with an 84 lb weight savings. Although P-DCPD has a lower tensile modulus than FRP, it is more flexible and better resists impact damage (564).

Reinforced RIM (RRIM) elastomers are also useful. By the addition of reinforcing fillers—milled glass fiber, glass flake, or mineral fillers in the PUR or short to long glass fiber in preforms, fabric, or mat forms placed in the mold cavity—the properties of the material can be altered to meet high performance requirements for the part (chapter 15). The reinforced elastomers are used to increase flexural modulus, improve thermal properties, and improve dimensional stability (563).

A probable first commercial use of a soy-based formulation in a high-density structural foam PUR RIM system is from John Deere, Bayer, and GI Plastek organizations, which launched a commercial program during 2001. Beginning with the 2002 model year, John Deere Harvester Works' entire line of combines included body panels molded with Harvest Form composite. One example is reviewed in Table 12.7.

PROCESS	DESIGN FLEXIBILITY	STRUCTURAL INTEGRITY	SECONDARY OPERATIONS	RELATIVE TOOLING COST	ASSEMBLY FLEXIBILITY
Reaction injectin molding	• Good design flexibility due to low pressures but large complex structural parts not feasible due to lower material properties • Thicker sections required • Batch process	• Lower properties can prohibit complex high-stressed features • Lower impact and creep resistance	• Flashing must be trimmed	• Low tooling cost	• Thermoset process such that thermal fastening techniques not possible
Structural foam molding	• Due to low pressures, significant design flexibility possible (parts consolidation, etc.) • High rigidity allows for high load-bearing structural members • No sink marks with integral function	• Good structural integrity • Low-molded-in stress provides low warp, dimensionally stable parts	• Sprue removal • Painting req. for appearance surfaces	• Lower tooling cost; aluminum tools possible	• Vibration welding, ultrasonic bonding, self-tapping screws, ultrasonic inserts, adhesive bonding possible • Many parts can be integrally molded
Injection molding	• some flexibility possible, but due to high pressures, large complex parts not cost effective • Ribs required for high load-bearing parts • Sink marks in thick sections	• Good structural integrity	• Sprue removal • Class A finish	• Higher pressures req. expensive steel tools: high-strength, prehardened	• In thermoplastics, vibration welding, ultrasonic bonding, self-tapping screws, ultrasonic inserts and adhesive bonding possible • Parts can be consolidated
Sheet molding compound	• Fiber orientation and resin-rich areas may occur in complex, load-bearing areas • Lower fatigue strength limits complex, dynamic parts • Limited deep draws on complex/large surfaces	• Possible nonuniform physical properties • Lower impact strength	• Deflashing • Large or small openings must be trimmed or cut out	• Steel tools required	• Thermoset materials: requires molded-in inserts
Sheet metal	• Only simple shapes and contours possible • Requires multiple dies for part complexity • Inferior dimensional control	• Minimal integral component strength due to multiple component assembly	• Multiple assembly operations: drilling, tapping, welding	• Low cost tooling • Complex deep-draw dies etc. • High piece part cost	• Screws, nuts, bolts, rivets, welding • Parts consolidation nearly impossible
Die casting	• Limited complex-part capability • Large parts are heavier	• Good structural integrity • Lower impact strength	• Trim dies required • Machining of critical surfaces	• High tooling cost • tool maintenance required due to potential wear damage	• Hardware assembly

Table 12.6 Structural foam information for large, complex products

OEM: John Deere

Material: Baydur® 730S IBS structural foam RIM system

Molder: G.I. Plastek

Weight: Approximately 75 pounds

Dimensions: Approximately 6-feet by 6-feet

Key properties: Lightweight, yet structurally stiff

Innovations: Commercial use of a soy-based polyurethane RIM formulation and use of in-mold coating technology

Table 12.7 John Deere rear shield made from a soy-based structural foam PUR RIM formulation

This durable, new composite is extremely strong, and yet it weighs 25% less than steel. Some Harvest Form panels utilize Bayer's Baydur structural foam PUR RIM system, which uses a soybean-based polyol component. This structural foam PUR RIM formulation is based on soybeans that produce physical properties and processing parameters equivalent to Bayer's conventional formulations.

One of the parts molded by GI Plastek using the soy-based Baydur material is the approximately 6 ft by 6 ft, 75-lb rear wall of the John Deere STS combine vehicle. GI Plastek adapted its proprietary ProTek in-mold coating system to the new material so that John Deere could continue to enjoy the benefits of this cost-saving alternative to painting.

There are other proprietary systems, such as polyacrylamate. It is Ashland Chemical's Airmax that is designed for use with preforms or glass mats. These reinforced plastics possess a high flexural modulus, good impact resistance, and high temperature stability. Systems with similar performance from isocyanate-based polymers are also used.

There are RIM systems based on chemistry unrelated to PURs that are not in significant commercial production compared to the PURs. Development work has taken place with these materials, such as nylon. The nylon RIM material is based on caprolactam. Nylon RIM polymers offer high toughness and abrasion resistance. Polydicyclopentadiene (P-DCPD) is a proprietary TS polymer developed by Hercules. P-PCPD offers high impact resistance and stiffness. It is used in the production of snowmobile components. Other polymers, such as epoxies, polyesters, acrylics, phenolics, and styrenics, are used.

Almost no other plastic has the range of properties of PUR. The modulus of elasticity range in bending is 200 to 1400 MPa (29000 to 203,000 psi), and it is heat resistant from 194°F to over

392°F (90°C to over 200°C). The higher values are for chopped glass fiber reinforcement added to the polyol blend that produces RIM (RRIM). The higher performance is obtained by injecting the polyol mix into a cavity with longer fiber constructions to produce structural RIM (SRIM). Figure 12.21 summarizes basic processes that produce RIM materials. An example of the density distribution across the thickness of a foamed part is shown in Figure 12.22 (565).

Conversion Process

In RIM, the starting point for the conversion process is liquid chemical components (monomers, not polymers). These components are metered out in proper ratio, mixed, and injected into a mold, where the finished product is formed. In reality, it is a chemical and molding operation combined into one system of molding in which the raw material is not a prepared compound but chemical ingredients that will form a compound when molded into a finished part. The chemicals are highly catalyzed to induce extremely fast reaction rates. The materials that lend themselves to the process are urethane, epoxy, polyester, and others that can be formulated to meet the process requirement. The system is composed of the following elements:

1. Chemical components that can be combined to produce a material of desired physical and environmental properties. Normally, this formulation consists of two liquid chemical components (three or more are also used) that have suitable additives and are supplied to the processor by chemical companies (three or more are also used).
2. A chemical processing setup that stores, meters, and mixes the components ready for introduction into the mold.
3. A molding arrangement consisting of a mold, a mold-release application system, and stripping accessories to facilitate smooth, continuous operation.

The success of the overall operation will depend on the processor's knowledge of (1) the chemistry of the two components and how to keep them in good working order; (2) how to keep the chemical adjunct in proper functioning condition so that the mixture entering the mold will produce the expected result; and (3) mold design, as well as the application of auxiliary facilities that will bring about ease of product removal and mold functioning within a reasonable cycle (such as 1 or 2 minutes).

The production of PUR involves the controlled polymerization of an isocyanate, a long-chain-backbone polyol, and a shorter-chain extender or cross-linker. The reaction rates can be controlled through the use of specific catalyst compounds, well known in the industry, to provide sufficient time to pour or otherwise transfer the mix and to cure the polymer sufficiently to allow handling of the freshly demolded part. The use of blowing agents allows the formation of a definite cellular core (thus the term "microcellular elastomer") as well as a nonporous skin, producing an integral sandwich-type cross section.

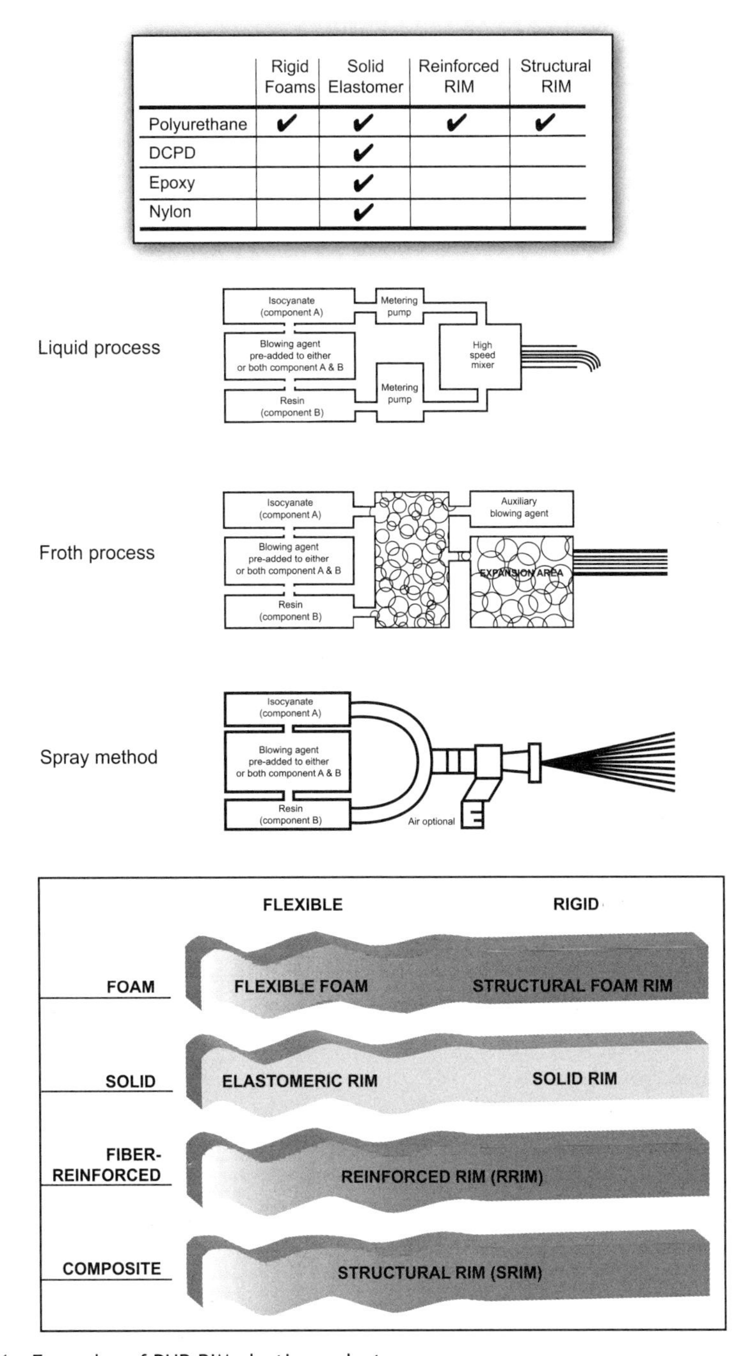

	Rigid Foams	Solid Elastomer	Reinforced RIM	Structural RIM
Polyurethane	✔	✔	✔	✔
DCPD		✔		
Epoxy		✔		
Nylon		✔		

Figure 12.21 Examples of PUR RIM plastic products.

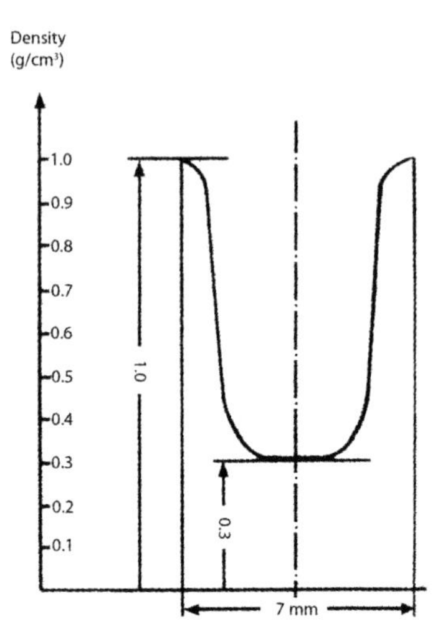

Figure 12.22 Density distribution across the thickness of a foamed part.

In RIM, all necessary reactive ingredients are contained in two liquid components: an isocyanate component and a resin component. The choice of isocyanate, as well as variations within isocyanate families, exerts a profound effect on the processing and final properties of the plastic. The chemical structures of two of the major diisocyanate types MDI and TDI, are commonly supplied in an 80/20 mixture of the 2,4- and 2,6-isomers. Early in the development of RIM systems, the MDI family was chosen over TDI, based on the following considerations:

1. *Reactivity*. Given the same set of coreactants, MDI and MDI types are more reactive than TDI. This can be used to advantage when short cycles are required.
2. *Available coreactants*. The high reactivity of the MDI types also makes available a larger number of coreactants. For example, where hindered aromatic amines yield a given level of reactivity, a variety of glycols can give equivalent reactivity, thus allowing more formulation versatility.
3. *Handling*. The MDI materials offer excellent handling characteristics due to comparatively low vapor pressure.
4. *Green strength*. The ortho-isocyanate groups of TDI are less reactive than the para-groups. Thus, at the end of the reaction that forms a polymer, the rate of reaction slows, resulting in green strength problems upon demolding. MDI does not suffer this deficiency.

As reviewed, RIM involves very accurate mixing and metering of two highly catalyzed liquid urethane components, polyol and isocyanate. The polyol component contains the polyether backbone, a chain extender or cross-linking agent, and a catalyst. A blowing agent is generally included in either the polyol or isocyanate component.

In order to achieve the optimum in physical properties and part appearance, rapid and homogeneous mixing is necessary. Insufficient or slow mixing results in either surface defects on the part or possible delamination or blistering during postcure.

The urethane liquid components are stored at a constant temperature in a dry-air or nitrogen environment. These components are delivered to high-pressure metering pumps or cylinders that dispense the respective materials at high pressure and accurate ratios to a mixing head. The materials are mixed by stream impingement. Additional mixing is generally encouraged via a static mixer (tortuous material path) incorporated into the runner system of the mold. Following the injection of the chemicals, the blowing agent expands the material to fill the mold.

The preferred route for high-volume RIM manufacturing is multiple clamps fed from a single metering pumping unit, the logic being that this is the most efficient way to utilize the capacity of the mold-filling equipment.

TP Polyurethane

The first PUR to become commercial, produced in 1937 by I.G. Farbenindustrie (which later became Bayer AG), was a TP. It was targeted to improve the properties of nylon. TP PURs are plastics that, after processing via heat and cooling into parts, are capable of being repeatedly softened by reheating.

TS Polyurethane

These TS plastics, after final processing into products, are substantially infusible and insoluble. They undergo a chemical reaction (cross-linking) by the actions of heat and pressure, oxidation, radiation, and/or other means, often in the presence of curing agents and catalysts. Curing actually occurs via polymerization and/or cross-linking. Cured TSs cannot be resoftened by heat. However, they can be granulated with the material being used as filler in TSs as well as TPs.

Cure of TS

Ideally, TS plastics should combine (1) low molecular weight during processing, to provide easy melt fluidity, and (2) infinite molecular weight in the end product, to provide maximum end-use properties. The organic polymer chemist uses myriad functional groups and reactions to produce this paradoxical combination of properties, with many of them in commercial use. Before considering them individually, however, it is best to start by noting that the molding of TS plastics has encountered a number of practical difficulties that have limited the rate of growth of this technology. These difficulties are based on conflicting requirements.

The process engineer, first of all, would like materials that have unlimited shelf life (warehouse storage before use), pot life (working time after the reactive components are mixed), and process working time in general (resistance to premature cross-linking between cycles, in dead spots, and during down time in general); in fact, the ideal is a one-part system in which a mixture of all the reactants would be stable indefinitely. All these requirements spell low reactivity.

On the other hand, once one has melt flowing into the mold, one would like the fastest possible reaction to produce final cure and a short process cycle for maximum process economy. This clearly means high reactivity.

Considering the total irreconcilability of these two conflicting demands, it is remarkable how far the ingenuity of organic polymer chemists has gone toward producing some reasonable compromises, and the range of balance in these compromises has increasingly diversified with general progress in the field. A variety of techniques are used:

1. Mixing of reactants as they are injected into the mold has been most highly developed in RIM technology.
2. Thermal activation can combine stability at low temperature with high reactivity at high temperature.
3. Mixtures of solids are stable at room temperature, but melt and cure rapidly at molding temperature.
4. Microencapsulation of the catalyst or curing agent can produce a stable one-part system that is activated by crushing or melting of the encapsulant during molding.
5. Latent catalysts are stable at room temperature but are liberated or otherwise activated at molding temperature.
6. "Blocked" reactants are stable at room temperature; at molding temperature they "unblock," liberating the reactant to permit cure. This is most commonly practiced in urethane.

A third difficulty in many TS systems is that they involve condensation reactions, which liberate gases, or volatile liquids that must be vented to permit production of solid and flawless parts. Venting is an established practice in molding of TSs.

Despite these difficulties, nearly all conventional TS plastics are potentially adaptable to molding. Table 12.8 provides information on chemical reactions.

POLYMERIZATION

Polymerization is basically the bonding of two or more monomers to produce polymers/plastics. It involves an addition or condensation reaction in which the molecules of a monomer are linked together to form large molecules whose molecular weights are multiples of that of the original substance, resulting in high-molecular-weight components.

C═C Double Bond. The reactivity of the C═C double bond is most commonly used in making the commodity thermoplastics—polyolefins, styrenics, and vinyls. But it is also used for thermoset cure of two commodity polymers—rubber and polyester—and also for smaller specialties such as EPDM 1,2-polybutadiene resins Hercules H-Resin DAP diallyl phthalate, bismaleimide, and some silicones. Some of these require catalysis, others occur spontaneously at molding temperature, and all are addition reactions that do not liberate any by-products. They are all conventionally practiced by compression molding. In theory all can be converted to injection molding, but commercial development is still quite small.

Aromatic C—H. In phenols (and anilines), the ortho and para C—H bonds are extremely activated. This reactivity is commonly used in the polymerization and cure of phenolic resins. Even though this cure produces water vapor, and often also formaldehyde and ammonia, phenolic molding powders have been successfully adapted to injection molding and represent one of the major penetrations of this type This could be a significant factor in making phenolics economically competitive with commodity thermoplastics while still delivering their superior thermoset end-properties—rigidity, heat resistance, and chemical resistance.

Epoxy. The —C—C— (with O bridging the two C atoms) group in epoxy "resins" contains acute angles that are much smaller than the normal positions of the electron pairs. This structure is therefore "strained" and unstable—reactive. And this reactivity permits easy cure by a variety of "hardeners," particularly active-hydrogen sources such as amines, acids, and alcohols, and even homopolymerization by acidic or basic catalysts. This high reactivity and easy cure is widely used in casting, impregnation, coatings, and adhesives. With the growth of interest in thermoset injection molding, clever formulators have devised a number of one-part systems using all the techniques described earlier, permitting small-scale specialty transfer and injection molding to become commercial realities.

Peroxides. Organic peroxides (R—O—O—R) and hydroperoxides (R—O—O—H) contain the O:O bond which is unstable and tends to separate into free radicals:

$$RO{:}OR \rightarrow RO^{\cdot} + {}^{\cdot}OR$$
$$RO{:}OH \rightarrow RO^{\cdot} + {}^{\cdot}OH$$

which in turn are very useful for initiating thermosetting cure reactions. These cure reactions fall into two distinct classes:

1. Addition polymerization of vinyl groups is used to cure unsaturated polyesters and DAP diallyl phthalate resins.

2. Hydrogen abstraction from saturated polymers:

$$RO^{\cdot} + \text{—}CH_2\text{—} \rightarrow ROH + \text{—}\dot{C}H\text{—}$$

produces unstable radicals on the polymer chains. As soon as such radicals appear on adjacent chains, they couple to produce cross-links:

$$\begin{matrix} \sim\sim\dot{C}H\sim\sim \\ + \\ \sim\sim\dot{C}H\sim\sim \end{matrix} \rightarrow \begin{matrix} \sim\sim CH\sim\sim \\ | \\ \sim\sim CH\sim\sim \end{matrix}$$

This technique is frequently used to cross-link polyethylene and to cure ethylene/propylene EPR rubber and silicone rubber.

To adapt the cure reaction to different polymers and different molding temperatures, the formulator can manipulate two primary variables:

1. The R group of the peroxide can be chosen from a wide variety of organic structures, giving a wide range of stability/instability and thus a wide range of processing temperatures.

2. The peroxide decomposition is activated not only by heat but also by addition of transi-

Table 12.8 Chemical reaction review

tion metal ions and reducing agents, giving an even wider range of processing temperatures and rates.

Given all these choices, peroxide cure has found a number of specialty uses in injection molding of thermosets.

O—H Group. The hydroxyl group is important in two types of thermosetting polymerization/cure reactions:

1. Reaction with isocyanate:

$$R{-}O{-}H + O{=}C{=}N{-}R' \rightarrow R{-}O{-}\overset{O}{\overset{\|}{C}}{-}\overset{H}{\overset{|}{N}}{-}R'$$

produces polyurethanes. This is an addition polymerization reaction widely used in RIM (reaction injection molding).

2. Polymerization and cure of urea, melamine, and phenolic resins is based on addition of formaldehyde to form the $—CH_2OH$ (methylol) group:

$$H{-}\overset{H}{\overset{|}{N}}{-}\overset{O}{\overset{\|}{C}}{-}\overset{H}{\overset{|}{N}}{-}H + CH_2O \rightarrow H{-}\overset{H}{\overset{|}{N}}{-}\overset{O}{\overset{\|}{C}}{-}\overset{H}{\overset{|}{N}}{-}CH_2OH$$

$$\underset{\underset{C}{|}}{H{-}N{-}H} \qquad \underset{\underset{C}{|}}{H{-}N{-}CH_2OH}$$

$$H{-}\overset{H}{\overset{|}{N}}{-}\overset{N}{\overset{\|}{C}} \quad \overset{N}{\overset{|}{C}}{-}\overset{H}{\overset{|}{N}}{-}H + CH_2O \rightarrow H{-}\overset{H}{\overset{|}{N}}{-}\overset{N}{\overset{\|}{C}} \quad \overset{N}{\overset{|}{C}}{-}\overset{H}{\overset{|}{N}}{-}H$$

$$N \qquad\qquad N$$

$$C_6H_5OH + CH_2O \rightarrow HOC_6H_4{-}CH_2OH$$

followed by further condensation of this methylol group with urea, melamine, and phenol and with other methylol groups. These reactions liberate water, and sometimes also formaldehyde, which must be vented carefully to prevent bubbles and cracks in the cured product; such venting techniques are conventional in compression molding, but are more difficult to adapt to conventional injection molding technique. Nevertheless, these reactions are fast, controllable, and capable of producing optimum thermoset properties with

Table 12.8 Chemical reaction review *(continued)*

improved process economics. Thus injection molding of phenolics has been quite widely adopted (22), and injection molding of urea resins has also been described (23).

N—H Group. Organic amines:

$$R{-}\ddot{N}(H)_2 \quad \text{(R—N: with two H)}$$

contain both the basic unshared electron pair on the nitrogen, which has strong catalytic effects; and also the reactive acidic hydrogens, which add readily to electron pairs in a wide variety of co-reactants. This provides the most common basis for cure of epoxy resins, the fastest and firmest cure of polyurethanes, and the basic mechanism for polymerization and cure of urea and melamine resins. For commercial development of thermoset injection molding, the cure of urethanes in RIM has been the most widely used, while the others have been used primarily in smaller specialty applications.

Isocyanate. The high reactivity of the isocyanate group with active hydrogen:

$$\underset{}{R{-}N{=}C{=}O} + \underset{\text{Alcohol}}{HOR'} \rightarrow \underset{\text{Urethane}}{R{-}N(H){-}C(=O){-}O{-}R'}$$

$$R{-}N{=}C{=}O + \underset{\text{Amine}}{H_2NR'} \rightarrow \underset{\text{Urea}}{R{-}N{-}C{-}N{-}R'}$$

$$R{-}N{=}C{=}O + \underset{\text{Water}}{HOH} \rightarrow \underset{\text{Amine}}{R{-}NH_2} + \underset{\text{Gas}}{CO_2}$$

is the basis for the success of polyurethanes in general and for the ease of conversion to RIM specifically. While most injection molding uses two-part systems which must be mixed immediately before injection into the mold, formulators have been applying their maximum ingenuity to the development of one-part systems based on all of the principles discussed earlier (24). Even among so-called thermoplastic urethanes, some of the best depend upon side-reactions that form allophanate and biuret cross-links in the hot mold to produce somewhat thermoset products with maximum end-use properties.

S—H Group. The active hydrogen of the mercaptan end-group in polysulfide rubber oligomers is generally used in two types of thermosetting systems:

1. Polysulfide elastomers are cured by oxidative coupling of the mercaptan groups:

$$2\,R{-}S{-}H + (O) \rightarrow R{-}S{-}S{-}R + H_2O$$

Evolution of water could thus be a problem to which the injection molder would have to adapt as he had to with phenolics.

2. Epoxy resins are sometimes cured with polysulfides in order to reduce their brittleness. Since this is a simple addition reaction, it should be readily adaptable to injection molding.

Table 12.8 Chemical reaction review *(continued)*

RRIM and Resin Transfer Molding

RRIM is very similar to resin transfer molding (RTM; chapter 14). In RRIM, process reinforcements, such as woven or nonwoven fabrics, short glass fibers, glass flakes, milled fibers, dry reinforcement preform, and so forth, are placed in a closed mold. Next, a reactive plastic system is mixed under high pressure in a specially designed mixing head. Upon mixing, the reacting liquid flows at low pressure through a runner system to fill the mold cavity, impregnating the reinforcement in the process. Once the mold cavity is filled, the plastic quickly completes its reaction. The complete cycle time required to produce a molded thick product can be as short as one minute.

These reinforcements provide stiffening or strengthening of the product and reduce thermal expansion. The usual procedure is to establish reinforcement in the mold cavity, using some type of clamping system, before the RIM process occurs. Milled fibers, such as glass, can be mixed into one of the liquid reactive material tanks, where a continuous stirring action exists.

The advantages of RRIM are similar to those listed for RTM (Fig. 12.23). However, RRIM can use preforms that are less complex in construction and lower in reinforcement content than those used in RTM. The RRIM plastic systems available will build up viscosity rapidly, resulting in a higher average viscosity during mold-filling. This action follows the initial filling with a low-viscosity plastic (565).

COSTING

Example of cost analysis is reviewed in Table 12.9.

Cost-to-performance advantages with RIM have been rather impressive. How RIM accomplished so much so quickly since 1969 is reviewed in a technical paper presented by Harry George of Bayer at the meeting of the Polyurethane Manufacturers Association, which took place between

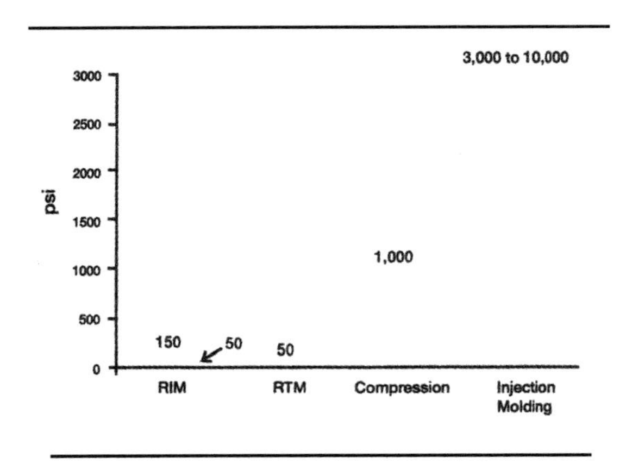

Figure 12.23 Molding pressure with RIM and RTM measures significantly less in other processes (courtesy of Bayer).

	PUR-RIM	Injection molding
Plastic temperature, °C	40–60	200–300
Plastic viscosity, Pa · s	0.5–1.5	100–1,000
Injection pressure, bar	100–200	700–800
Injection time, s	0.5–1.5	5–8
Mold cavity pressure, bar	10–30	300–700
Gates	1	2–12
Clamping force, t	80–400	2,500–10,000
Mold temperature, °C	50–70	50–80
Time in mold, s	20–30	30–80
Wall/thickness ratio	1/0.8	1/0.3
Part thickness, typical maximum, cm	10	1
Shrinkage, %		
Unreinforced	1.30–1.60	0.75–2.00
Reinforced—glass		
parallel to fiber	0.25	0.20
vertical to fiber	1.20	0.40
Inserts	Easy	Costly
Mold alterations	Cost-effective	Costly

Cost includes: (1) raw material, (2) finished part inventory, (3) capital cost of equipment that includes auxiliary and handling equipment, (4) equipment reliability as it relates to output rate, (5) cycle time vs. production schedule, and (6) availability of spare parts.

Table 12.9 Example of cost analysis of PUR RIM and injection molding of products with large surface areas

September 30 and October 1, 2002. It provides a detailed look at the advantages, chemical formulations, and materials and processing innovations that make RIM technology well suited to a wide range of applications in diverse markets, including automobile manufacturing, industrial components, recreational products, and enclosures for sophisticated medical diagnostic equipment. The latest news about RIM and its applications is available at BaySystems's Web site, http://www.Bayer-BaySystems.com.

CHAPTER 13

ROTATIONAL MOLDING

INTRODUCTION

Rotational molding (RM) is also called rotomolding, rotational casting, centrifugal casting, or corotational molding. This method, like blow molding (chapter 6) and thermoforming (chapter 7), is used to make hollow one-piece parts (Table 13.1). The process is based on the heating and cooling of an axially or biaxially rotating split-hollow cavity mold that defines the outside shape of the required product. No pressure is applied, other than the relatively low-contact pressure developed during the rotation of the heated melts. The most common is the multiarm turret machine that has a three-stage operation (13, 104, 105, 151, 153, 333, 356, 538, 544–549, 551, 553–555).

A premeasured amount of powder or liquid thermoplastic (TP) material is placed in the cavity, which is mounted on a turret arm capable of rotating the mold. This action permits a uniform distribution of the plastics that is forced against the inside surface of the cavity. Following a prescribed cycle, the heat of the oven fuses or sinters the plastic and goes into the cooling chamber. The solidified product is removed from the mold and the cycle is repeated. This process permits molding very small to very large products. To improve product properties, hasten product densification, reduce air voids, reduce cure time, and so on.

The cycle times typically range from 6 to 12 minutes. They can be at least 30 minutes for large parts. The wall thickness of the parts affects cycle times, but not in a direct ratio. For example, with polyethylene (PE) plastic, the cycle time increases by about 30 seconds for every 25 mils of added thickness up to ¼ in thickness. Beyond ¼ in, the heat-insulating effect of the walls increases cycle times disproportionate to any further increase in thickness; cycle times usually have to be determined experimentally or with prior experience.

The mold in the oven spins biaxially with rotational speeds being infinitely variable, usually ranging up to 40 rpm on the minor axes and 12 rpm on the major axes. A 4:1 rotation ratio is

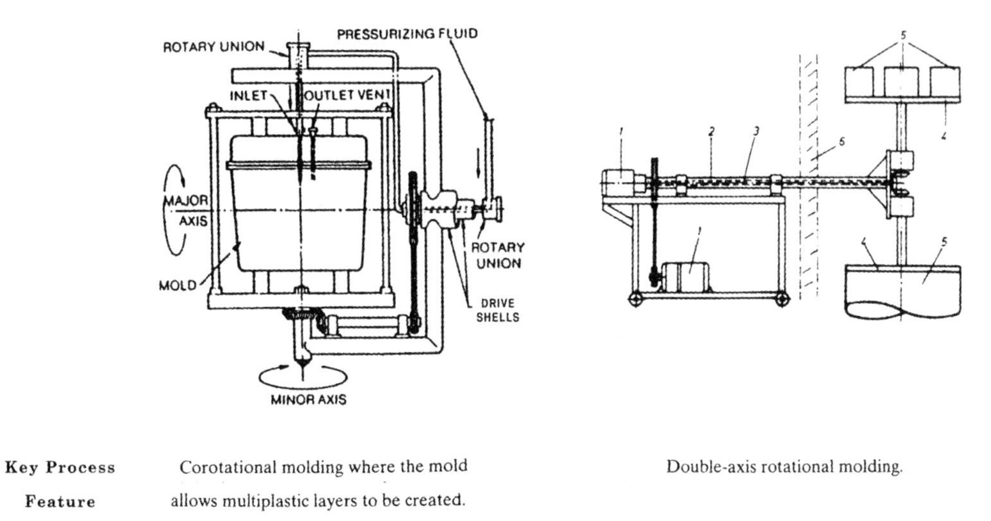

Key Process Feature	Corotational molding where the mold allows multiplastic layers to be created.	Double-axis rotational molding.

Table 13.1 Comparison of different processes

generally used for symmetrically shaped parts. A wide variety of ratios are necessary for molding unusual and complex shapes.

Venting molds are often used to maintain atmospheric pressure inside the closed mold during the entire molding cycle. A vent will reduce flash and prevent mold distortion as well as lower the pressure needed in the mold to keep the mold closed. It will prevent blowouts caused by pressure and permit use of thinner molds. For example, the vent can be a thin-walled plastic tube of polytetrafluoroethylene (PTFE) that extends to near the center of the cavity. It enters the mold at a point where the opening it leaves will not affect the parts' appearance and other things. The vent can be filled with glass wool to keep the powder charge from entering the vent during rotation.

Products that employ RM in their production include furniture, light shades, marine accessories, material-handling bins, shipping drums, storage tanks and receptacles, surfboards, toys, and so on. End-use product sizes range from small balls up to at least 22,000 gallon tanks (83 m^3) that weigh at least 2½ tons (5000 lb).

PROCESS

The four-step RM process essentially consists of charging a measured amount of plastic (basically the weight of the molded product) into a mold that is rotated at relatively low speeds (usually 4 to 20 rotations/min), usually in an oven (electrical heating has also been used more recently), around two perpendicular axes (Fig. 13.1). In the oven, the heat penetrates the mold, causing the plastic, if it is in powder or granular form, to become tacky (Tables 13.2 and 13.2) and stick to the mold's female cavity surface; or if it is in viscous liquid form, the plastic will start to gel on the mold

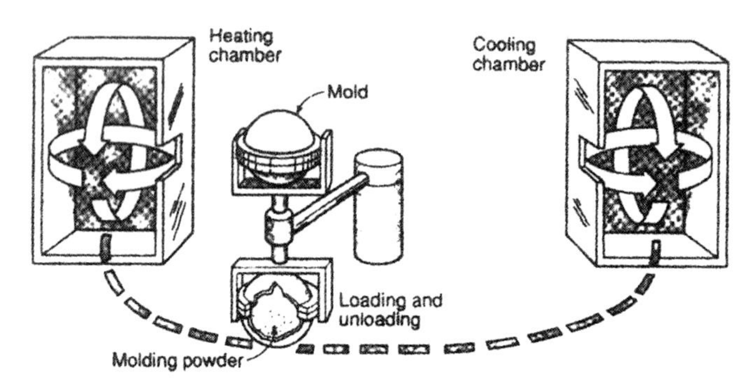

Figure 13.1 RM's four basic steps (courtesy of The Queen's University, Belfast).

Polymer	Melt Temperature, °C	Glass Transition Temperature, °C	Tack Temperature, °C*	Kink Temperature, °C
LDPE	120±1	—	115±5	NA
MDPE	125±5	—	120±5	100
HDPE	130±1	—	130±5	NA
PP	165±5	—	155±5	120
Nylon 6	225	—	NA	175
APET	—	80	100±5	110
GPPS	—	105	110±5	NA
MIPS	—	105	120±5	NA
ABS	—	105	125±5	117
PMMA	—	105	105±5	NA
PC	—	155	160±5	155

* Measured by blowing -35 mesh polymer powder against a hot plate held in a vertical position

Table 13.2 Tack temperatures for different plastics

surface. Solid pellets of plastic are required to be rather smaller and more uniform than the type used in other processes, such as injection molding (105).

Since the molds continue to rotate while the heating continues, the plastic will gradually become distributed relatively evenly on the mold female cavity walls through gravitational force. As the cycle continues, the plastic melts completely, forming a homogeneous layer of molten plastic. The mold is cooled following the heating step. With the mold continuing its rotation, air from a

T_{oven} (°C)	Relative Time to Reach a Tack Temperature of 100°C	Relative Time to Reach a Tack Temperature of 125°C
275	1.12	1.58
300	**1.0**	1.40
325	0.9	1.25
350	0.82	1.14
375	0.76	1.04
400	0.7	0.96

Table 13.3 Relative time to reach two tack temperatures at different oven temperatures

high-velocity fan or a fine spray of water (sometimes both) cools the mold. After the melted plastic cools, the molds are opened and the parts are removed (Tables 13.4 and 13.5; 104, 105, 151).

RM pressure and temperatures are unique compared to most other manufacturing processes because no mechanical/pressure shearing heat is used to melt the plastic. This process relies on convection heat (oven-mold-plastic) to melt the plastic. Starting temperatures depend on the plastic being processed. For example, an oven's starting temperature can be from 500°F to 600°F (260°C to 316°C). A common mistake is to use excessively high temperatures to shorten the cycle time (Table 13.6). An important consideration for consistency is to check the temperature of the mold before it goes into the oven and account for any variation in the temperature. Environmental conditions, along with the location of the mold storage, may cause the variation. Machines usually vary in

Fluid	Convection Heat Transfer Coefficient,	
	$\times 10^{-3}$ W/cm^2 °C	Btu/ft^2 hr °F
Quiescent air	0.5 – 1	0.8 – 2
Air moved with fans	1 – 3	2 – 5
Air moved with blowers	3 – 10	5 – 20
Direct combustion gas impingement	6 – 10	10 – 20
Air and water mist	30 – 60	50 – 100
Fog	30 – 60	50 – 100
Water spray	30 – 90	50 – 150
Oil in pipes	30 – 180	50 – 300
Water in pipes	60 – 600	100 – 1,000
Steam in pipes, condensing	600 – 3,000	1,000 – 15,000

Table 13.4 Heat transfer coefficients during mold cooling

Steps
Predetermine the amount a plastic material that is based on the weight of the product to be molded.
If the mold had been used check that the mating surfaces and vent for cleanliness.
Put the plastic in one-half of the mold cavity (called "shot size" or "charging" the mold).
Close the mold halves and clamping them together.
Start the biaxial rotation and place the mold into the oven for the predetermined heat cycle. Rotation is around two different axes that are called the primary and secondary axes.
The ratio is dependent on the mold's shape and design but is typically 4:1 for simple shapes or even 8:1 depending on the complexity of the product.
Speeds are usually <20-rpm. Low melt index plastics require slower speeds and higher melt index plastics require higher speeds.
As the plastic is heated and melts it adheres to the inner surface cavity of the mold. The process continues until the entire shot size has formed a coating within the mold. During heating the plastic material melts, fuses, and densifies into the shapes of the internal cavity by the controlled directional centrifugal forces.
The mold then goes to the cooling cycle while continuing to rotate. Occasionally, this occurs in the oven but most often occurs at a separate cooling station.
Cooling can be accomplished with air from a high-velocity fan which is relatively slow and limits molded-in stress, or more commonly with water which is faster than air but can cause molded-in stress.
After cooling and the mold is opened the finished solid product is removed.
Occasionally products are placed into a fixture to maintain dimensional stability while cooling.

Table 13.5 Steps taken during the RM fabrication process

Characteristic	Length of Oven Time						
	Very Short	**Short**	**Almost Right**	**Optimum**	**Slightly over Optimum**	**Longer than Needed**	**Excessive**
Odor	None	Little	Somewhat waxy	Waxy	Pungent	Very acrid	Burnt
Inside surface color	⟵ Same as outside surface ⟶				Slightly yellow	⟵Increasing to brown⟶	
Inside surface appearance	⟵ Dull, matte ⟶			⟵ Shiny, glossy ⟶			
Inside surface	Very rough texture	Rough	Waxy	Not sticky	Smooth, slightly sticky	Sticky	Very sticky
Inside bubbles	Very many	Many	Few to none	⟵ None ⟶		Few	Gross
Outside bubbles	Many	Few	Few to none	⟵ None ⟶		Few to many	Many
Fill	Bridging	⟵ Some ⟶		⟵ Complete ⟶			
Tear resistance	Poor	⟵Better⟶		⟵ Maximum ⟶			Decreasing

Table 13.6 Effect of oven heat time on RM plastics

the placement of sensors, accuracy of controlling temperatures, and number of instruments used for monitoring. If these variables exist it can become extremely difficult to repeat processing conditions between cycles or machines.

The temperature and duration of the heating cycle needs to be controlled. Most machines that are being built have horizontal rotating arms with closed, recirculating, high-velocity, hot-air ovens with total automation of the complete process. Many of these machines are computers programmed to obtain consistent product quality.

The mold in the oven spins biaxially with rotational speeds being infinitely variable, usually ranging up to 40 rpm on the minor axes and 12 rpm on the major axes. A 4:1 rotation ratio generally is used for symmetrically shaped parts (Table 13.7). A wide variety of ratios are necessary for molding unusual and complex shapes, such as boats (Fig. 13.2). Actual ratio relates to the melt flow characteristics of the plastic being processed.

PLASTIC

Nearly all RM products are made from TPs, although thermoset (TS) plastics can be used. Figure 13.3 provides a guide to the types of plastics used. Linear, low-density PE is the major plastic

Speed Ratio	Shapes
8:1	Oblongs, straight tubes (mounted horizontally)
5:1	Ducts
4:1	Cubes, balls, rectangular boxes, most regular 3-D shapes
2:1	Rings, tires, mannequins, flat shapes
1:2	Parts that show thinning when run at 2:1
1:3	Flat rectangles, suitcase shapes
1:4	Curved ducts, pipe angles, parts that show thinning at 4:1
1:5	Vertically mounted cylinders

Table 13.7 Examples of rotational ratios for different shapes

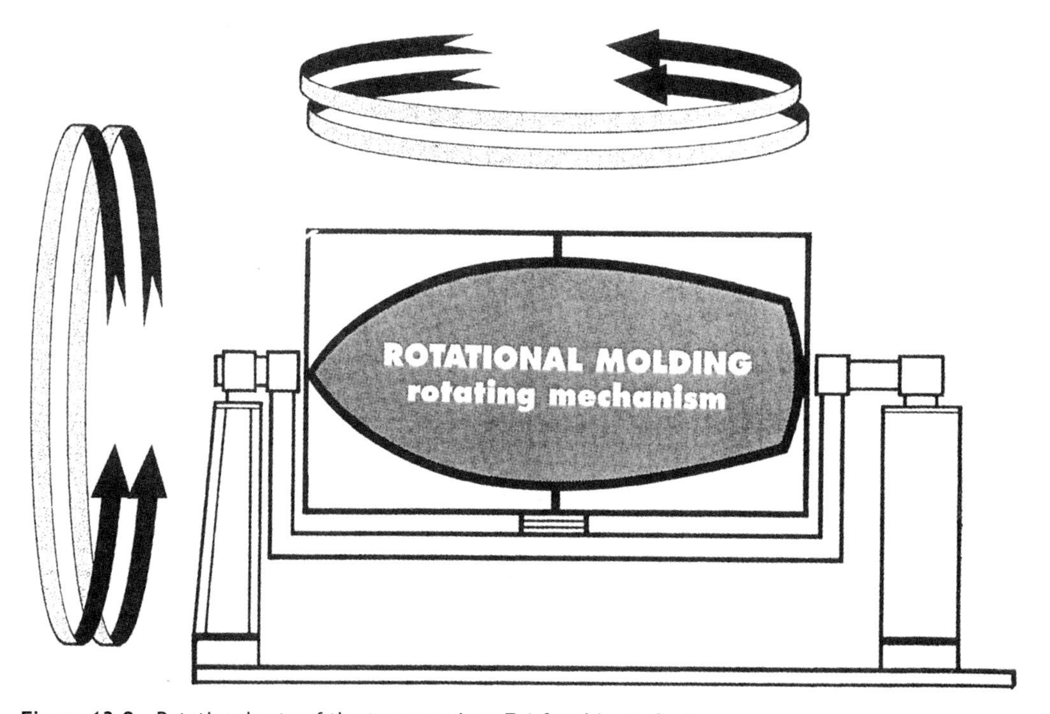

Figure 13.2 Rotational rate of the two axes is at 7:1 for this product.

used, with about 85wt% of all plastics representing different forms of PEs (557). Other plastics include nylon, polycarbonate (PC), TP polyester, and polypropylene (PP; 105). These plastics can be solid or foamed (Table 13.8; chapter 8 reviews chemical blowing agents [CBAs]). Table 13.9 and Figures 13.4 to 13.6 provide examples of RM products that range from a small PE beach ball to a cross-linked PE (XLPE) 22500-gallon tank. This size of tank, with a 1½ in wall, uses a triple XLPE charge; the first charge is about a 2500 lb, and the following two measure 1500 lb each.

The plastic powder form usually has a particle size of 35 mesh (74 to 2000 μm). The other form is liquid. Some high-flow plastics, such as nylon, have been used in small pellet form. Ethylene vinyl acetate and PEs are also used in specialized applications, as are polyvinyl chloride (PVC), PC, TP polyester, nylon, and PP. RM vinyl plastisols produce beach balls, floating animals, and toys, as well as industrial products (Table 13.10). The liquid or powdered plastic used in this method flows freely into corners or other deep draws upon the rotation of the mold and is then fused by heat passing through the mold's wall.

The particle size of powders is usually quantified in terms of the mesh size. This relates to the number of mesh openings per inch in the sieve used to grade the powder. Table 13.11 gives some of the mesh sizes defined in the American (ASTM E-11) and British standards. Particle shapes vary (Table 13.12). The aim is to use particles that have low packing density (Table 13.13). Table 13.14 compares powders with micropellets. The way powder flows in a mold during heat processing is important to the performance and aesthetics of products (Table 13.15). Packing the powder as

CBA Level (% wt)	CBA[1] Type	Oven Temperature (°C)	Oven Time (min)	Comments
1	OBSH	246	10	Good inside skin, limited foaming
1	OBSH	246	12	Good inside skin, good foam
1	OBSH	246	14	Fair inside skin, good foam
1	AZ	260	10	Good inside skin, little foam
1	AZ	260	12	Good inside skin, good foam
1	AZ	260	14	Poor inside skin, overblown with coarse cells

1. CBAs (chemical blowing agents reviewed in Chapter 8).
(OBSH = *p,p′*-oxybisbenzene sulfonyl hydrazide; AZ = azodicarbonamide)

Table 13.8 Effect of oven condition on foaming high-density PE (HDPE)

Tanks		
Industrial tanks	Septic tanks	Chemical storage tanks
Nonplastic tank liner	Oil tanks	Fuel tanks
Agricultural equipment	Water treatment tanks	Shipping tanks
Automotive		
Bumper	Door armrests	Instrument panels
External 3-D panels	Traffic signs/barriers	Ducting
Wheel arches	Fuel tanks	
Containers		
Refrigerated boxes	Reusable shipping containers	
Planters	Airline containers	
Drum/barriers	Refuse containers	
Toys and Leisure		
Sporting goods	Playhouses	Doll heads and body parts
Outdoor furniture	Balls	
Hobby horses	Ride-on toys	
Materials Handling		
Skids	Pallets	Fish bins
Packaging	Trash cans	
Storage bins	Carrying cases for paramedics	
Marine Industry		
Dock floats	Leisure craft/boats	
Pool liners	Kayaks	
Docking fenders	Life belts	
Miscellaneous		
Lawn equipment	Manhole covers	Tool boxes
Garden equipment	Housings for cleaning equipment	Dental chairs
	Point-of-sale advertising	Consumer products

Table 13.9 Examples of RM products

tight as possible is essential to help obtain the densest finished molded product. Spherical particles should be avoided because they do not pack as well as those that are close to being rectangular or square (105).

The plastics used in RM are generally more expensive than the pellet plastics used in many other processes, because they must be more finely and evenly powdered. A degree of compensation on plastic used in RM exists. The process generates low or no levels of regrind or scrap, even when it is operating poorly. Products can have no flash at all if properly designed molds are used.

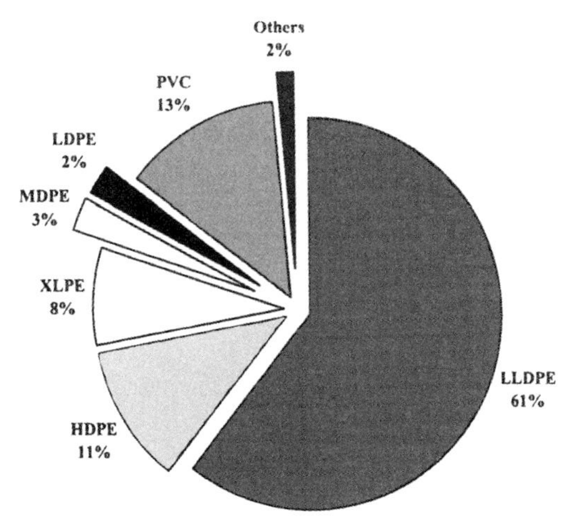

Figure 13.3 Consumption of plastics for RM.

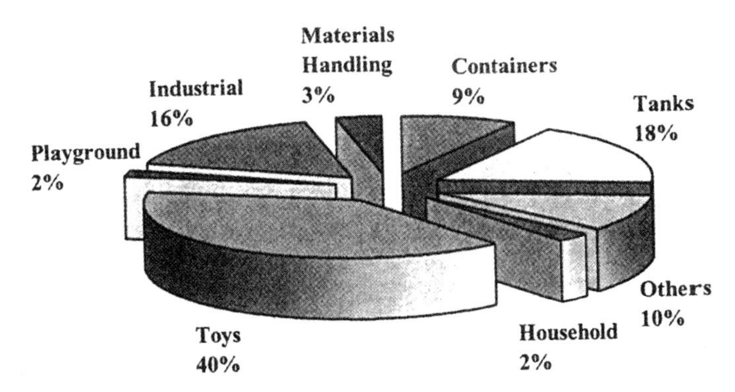

Figure 13.4 RM products in North America.

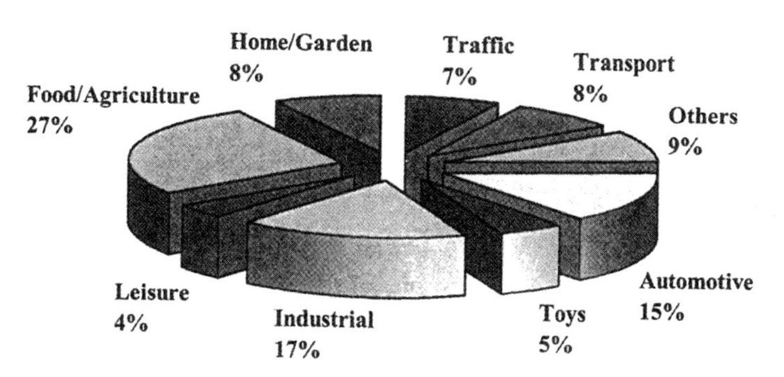

Figure 13.5 RM products in Europe.

Figure 13.6 Example of RM products including large tank.

Condition	Plastisol	Drysol	Micropellet
State	Liquid	Dry powder	Micropellet
Dispensing	Liquid pump	Weigh-and-dump	Weigh-and-dump
Ease of dispensing	Moderate	Easy	Easy
Dispensing problem	Slop	Dusty	Little
Clean-up	Difficult, scraping	Moderately difficult	Moderate

Table 13.10 Examples of PVC plastics used in RM

Tyler Size	Sieve Opening (× 0.001 inch)	Wire Diameter (× 0.001 inch)	Particle Size (microns, μm)
35	16.5	11.4	420
60	9.8	7.1	250
80	7.0	5.2	177
100	5.9	4.3	149
115	4.9	3.6	125
150	4.1	3.0	105
170	3.5	2.5	88
200	2.9	2.1	74
250	2.5	1.7	63
325	1.7	1.2	44
400	1.5	1.0	37

Table 13.11 Sieve sizes

Average thickness	The average diameter between the upper and lower surfaces of a particle at its most stable position of rest.
Average length	The average diameter of the longest chords measured along the upper surface of a particle in the position of rest.
Average breadth	The average diameter at right angles to the diameter of average length along the upper surface of a particle in its position of rest.
Chunkiness	Reciprocal of elongational ratio.
Circularity	Ratio of the circumference of a circle with the same projected area to the actual circumference of the projected area.
Elongational ratio	The largest particle length to its largest breadth when the particle is in a position of rest.
External compactness	The square of the diameter of equal area to that of the profile, divided by the square of the diameter of an embracing circle.
Feret's diameter	The diameter between the tangents at right angles to the direction of scan, which touch the two extremities of the particle profile in its position of rest.
Martin's diameter	The diameter which divides the particle profile into two equal areas measured in the direction of scan when the particle is in a position of rest.

Table 13.12 Classifying particle shape for irregular particles

Projected area diameter	The diameter of a sphere having the same projected area as the particle profile in the position of rest.
Roundness factor	Ratio of the radius of the sharpest corner to the most round corner with the particle in a position of rest.
Specific surface diameter	The diameter of the sphere having the same ratio of external surface area to volume as the particle.
Surface diameter	The diameter of the sphere having the same surface area as the particle.
Stokes diameter	The diameter of the sphere having the same terminal velocity as the particle.
Volume diameter	The diameter of the sphere having the same volume as the particle.

Table 13.12 Classifying particle shape for irregular particles *(continued)*

Polymer	Compact Density (kg/m^3)	Reduced Density	Bulk Density	
			(kg/m^3)	(lb/ft^3)
LLDPE	910	0.38 – 0.43	345 to 390	22 to 24
HDPE	960	0.35 – 0.50	335 to 480	23 to 30
PS	1050	0.30 – 0.55	315 to 580	22 to 36
PP	910	0.25 – 0.40	230 to 365	14 to 23
Nylon	1100	0.40 – 0.60	440 to 660	27 to 41
FEP	2200	0.25 – 0.40	550 to 880	34 to 55

Table 13.13 Typical powder bulk density

RM is one of the fastest-growing sectors of the plastics processing industry, with annual growth rates (not during recessions) in the range of 10 wt% to 12 wt%. While RM has many advantages—low mold costs, seamless and stress-free moldings, controlled wall thickness distribution, among others—it is characterized by its slow cooling rates and lack of shear during shaping. These two factors lead to unique structural features in RM products. The absence of shear does not encourage good mixing of additives such as pigments, and the slow cooling promotes large spherulitic growth. On top of this, the longer cycle times and the presence of oxygen at the inner free surface of the moldings means that degradation processes can be initiated very quickly if the correct molding conditions are not used.

Effect	Micropellets	Powders
Particle size distribution	Very narrow (300–500 microns)	35 to 200 mesh (75–400 microns)
Cycle time	Extended	Normal
Porosity	Can be a problem	Normal
Color plate-out or staining	Moderate to low	Moderate to severe
Airborne dust	Low	Can be a nuisance
Color changeover	Recompound, slow	Dry-blend, fast
Color dispersion	Consistent	Can be a problem with certain dry-blending colors
Source of raw material	Extrusion	Extrusion + pulverizing
Pulverizing cost	None	\$0.06/lb to \$0.15/lb or so
Extrusion cost	Owing to lower throughput, perhaps \$0.05/lb to \$0.15/lb	None

Table 13.14 Comparing powders with micropellets

PLASTIC BEHAVIOR

The RM process has characteristic features that make the microstructure of the molded plastic articles unique. The fact that the rotation speeds are slow results in low shear, which promotes the development of textures that are orientation-free but prevents the dispersion of additives such as pigments. If the heating of the plastic is too severe, degradation may occur at the inner free surface, causing the spherulites that grow freely at that surface to be replaced by a nonspherulitic or trans-crystalline texture, depending on the extent of degradation (chapter 1). The plastic microstructure in the bulk, and also at the inner surface layer, has a major influence on the mechanical properties of the molded material (104).

PE and PP RMs influence a number of factors, such as molding temperature, grinding and mixing conditions, type of pigment, antioxidant level, mold material, and inner atmosphere. Test programs have found that PE and PP show different degradation behavior. While the thermo-oxidative degradation causes mainly chain scission in PP, cross-linking dominates in PE (560). The use of increased amounts of antioxidant in the plastic, or the use of an inert atmosphere, delays the degradation but does not prevent it. It has also been observed that the thermomechanical effects caused by the mixing processes commonly used to add pigments increase the mechanical properties of PE products. The nucleating activity of the pigment, combined with the mixing process, has a major effect on the microstructure and the mechanical properties of the final products.

Type	Comment
Steady-state circulation	Ideal flow
	Maximum mixing
	Best heat transfer
	Spherical or squared egg particle shape
	Cohesive-free or freely flowing powders
	Smooth powder surfaces
	Relatively high friction between mold surface and powder bed
Avalanche	Adequate powder flow
	Relatively good powder mixing
	Relatively good heat transfer
	Squared egg, acicular, or disk-like particles
	High friction between mold surface and powder bed
Slip flow	Poor powder flow
	No powder mixing
	Poor heat transfer
	Disk-like, acicular particles
	Powders with high adhesion or cohesion
	Agglomerating or sticky powders
	Very low friction between mold surface and powder bed

Table 13.15 Types of powder flow

While there has been extensive work done on the mechanical properties of RM products and the heat transfer processes occurring during molding, there has been relatively little work done on the unique microstructures that occur within RM products. This is surprising in view of the fact that the structures are very amenable to microstructural analysis. They display classic features in terms of spherulite geometries, and it is possible to relate mechanical properties to defects caused by the onset of thermal/oxidative degradation.

The effects of grinding at three temperatures (40°C, 60°C, and 80°C), as well as mixing by turbo blending and extrusion of pigments having different nucleating capabilities, was also investigated.

The moldings were characterized by several techniques: optical microscopy, differential scanning calorimetry, Fourier-transform infrared spectroscopy (FTIR), tensile and impact testing, and cone-and-plate oscillatory rheometry (chapter 22).

EFFECT OF THE THERMAL TREATMENT

The microstructure and the properties of plastics are very sensitive to the thermal treatment given to them during the molding stage. Measuring the maximum temperature of the atmosphere inside the mold during RM permits the thermal history of the plastic to be monitored and related to structural variations. Polarized light microscopy of cross-sections of RM products shows that when this temperature is too low, the samples have voids and the crystalline texture is spherulitic all through the thickness (173).

If the heating is too severe and the temperature inside the mold goes above a certain value, the morphology changes at the free inner surface. In the case of PE, the changes in microstructure as the overheating increases are very clear. For moderate overheating, the size and perfection of the spherulites near the inside surface reduces, and an inward-growing layer having columnar structure starts to appear (173).

When the heating is too severe, the spherulitic texture disappears completely near the inside surface and is replaced by a dark ribbon of material, apparently without any texture. For PP, the overheating of the plastic does not prevent the crystallization of spherulites near the inner surface but increases their size and birefringence. The increase in birefringence suggests that the lamellae cross-hatching of the type I spherulites might have reduced.

Fluorescence microscopy, which enables the detection of double bonds in polymer chains, showed that the altered microstructure at the inner surface fluoresces, indicating that the thermo-oxidative reactions took place there during the heating of the plastics. Further evidence of degradation, when overheating occurred, was given by FTIR spectra obtained from layers cut near the inner skin. Carbonyl, vinyl, and hydroperoxide groups that are common products from oxidative reactions of polyolefins were identified in the spectra. The use of a nitrogen atmosphere and an increase in the concentration of antioxidant did not prevent these morphologies from forming but raised the temperature at which they occurred.

The mechanical properties of the samples are very sensitive to the maximum inner air temperature (Fig. 13.7). The best properties were obtained with the samples that were heated up to a temperature that facilitated the diffusion of air out of the melt without causing degradation at the inner surface. The optimum inner air temperature varies with a number of factors. The use of nitrogen or a higher concentration of antioxidant delayed the degradation process and shifted the optimum temperature to higher values (Fig. 13.7[a]).

A similar effect is observed when the heating rate increases (Fig. 13.8). This effect is associated with the time of exposure of the plastic to high temperatures that decrease when the heating rate is increased. The temperature of grinding of the plastic before molding also affects the optimum process temperature (Fig. 13.9). A grinding temperature of 176°F (80°C) helps the plastic

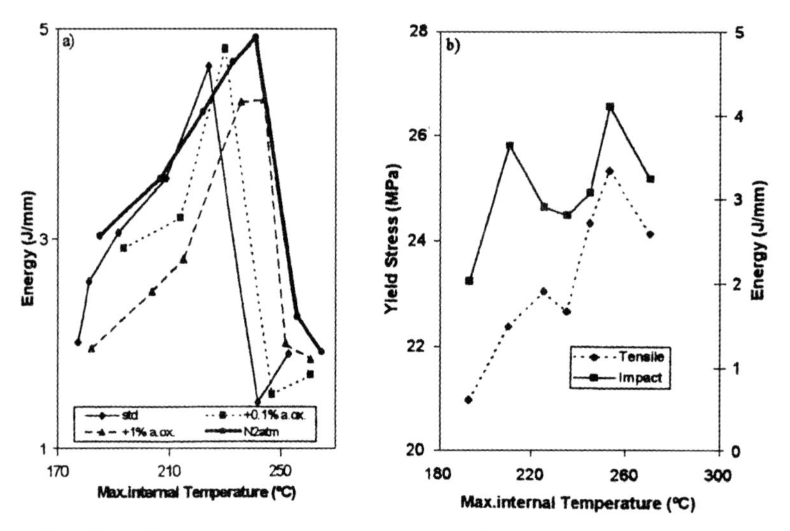

Figure 13.7 The effect of maximum inner temperature on the impact strength of the moldings (a = PE and b = PP).

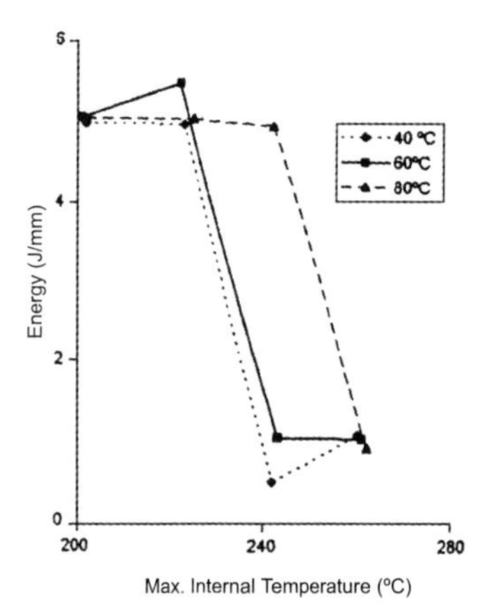

Figure 13.8 Effect of heating rate on the optimum processing temperature of PE.

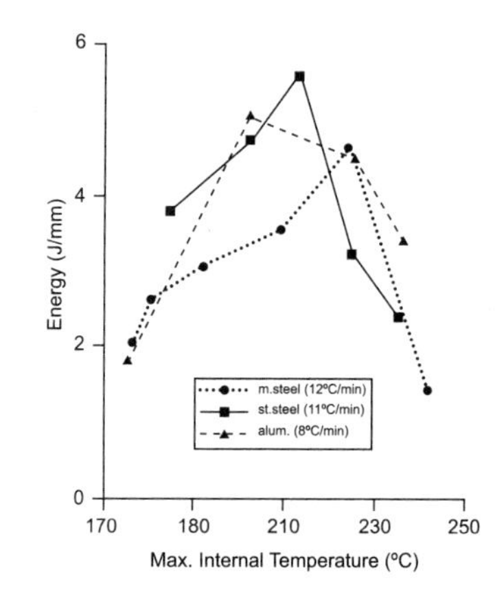

Figure 13.9 Effect of the grinding temperature on the optimum processing temperature of PE.

withstand higher processing temperatures (467°F or 242°C) without losing impact strength. The reason for this behavior may be the annealing effect of the grinding at a higher temperature, which increased the crystallinity and consequently the amount of energy absorbed by the polymer during the melting.

EFFECT OF PIGMENTATION AND MIXING METHOD

The turboblending of the plastic will further change the plastic's crystallization behavior. It is observed that, in the absence of pigment, rows of small spherulites grow at the plastic particle's boundaries. This might be due to the thermomechanical action of the rotating blades, which somehow modifies the structure of the molecules around the particles. The extrusion mixing reduced the size of the spherulites, probably due to an orientation memory of the extruded material that induced nucleation during the molding stage.

Polarized light and bright field micrographs are used to illustrate the effect of the nucleating ability and mixing efficiency of the pigments into the microstructure of PE samples. The nucleating effect is more noticeable in the turboblended samples, because as the pigment is spread around the original polymer particles, transcrystalline textures were formed at their boundaries. The nucleating properties of the pigment also have an effect on the microstructure of the extruded samples, although the mechanism of this is not yet fully understood.

It is apparent from the differential scanning calorimetry thermograms shown in Figure 13.10 that the extrusion process prior to molding increases the enthalpy of fusion and the crystallization temperature of the plastic (chapter 22). As expected from other works with nucleating additives, a similar increase was observed when a nucleating pigment was added by turboblending (Fig. 13.11).

It is an important observation that the microstructural modifications caused by the extrusion and turboblending in the unpigmented samples result in higher impact strengths than those observed in the original polymer (Fig. 13.12). This improvement in properties disappears if a pigment is mixed by turboblending, probably due to the poor distribution capability of this technique.

CONCLUSION

In this work, the microstructure of RM samples has been related to the processing conditions and mixing of additives. This helped to explain the variations in mechanical properties. The following main conclusions can be drawn:

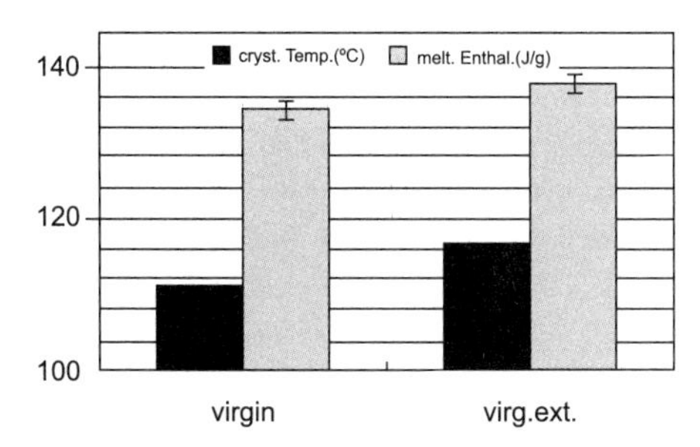

Figure 13.10 Effect of extrusion on the thermal properties of PE.

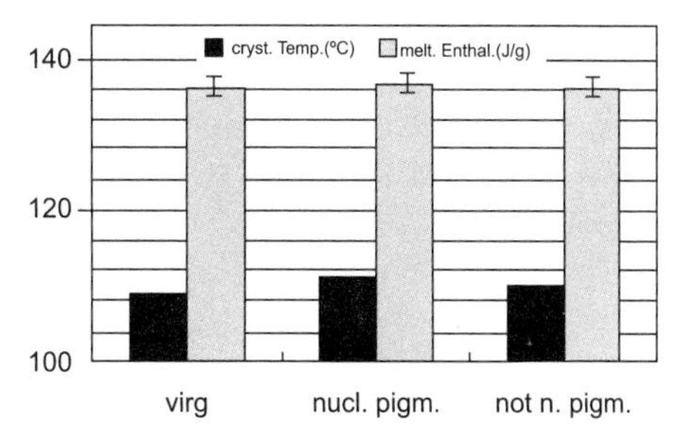

Figure 13.11 Effect of pigmentation on the thermal properties of turboblended PE.

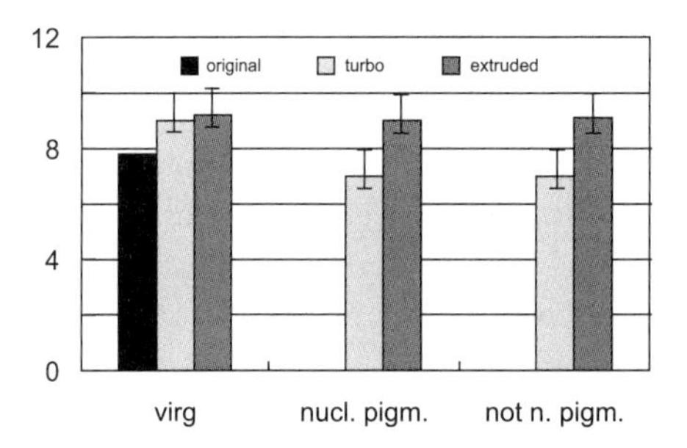

Figure 13.12 Effect of pigmentation and mixing on the impact strength of PE.

1. The maximum temperature reached by the air inside the mold is the more adequate parameter to control the process. It has a clear relationship with the microstructure and the properties of the moldings. Optimum values of these properties correspond to moldings free from voids and from degradation at the inner surface.
2. In RM, PE and PP show a different degradation behavior. The thermo-oxidative degradation mainly causes chain scission in PP, but cross-linking dominates in PE.
3. The use of increased amounts of antioxidant, an inert atmosphere, or a higher heating rate can delay the degradation. Grinding the plastic at a higher temperature (80°C) seems to have a similar effect.
4. The thermomechanical action of the mixing processes, commonly used to add pigments, affects the microstructure and improves the mechanical properties of PE moldings.
5. The nucleating activity of the pigment, combined with the mixing process, has a major influence on the microstructure and properties of the final products.

PERFORMANCE

The rigidity or flexibility of the molded product is controlled by the formulation of the plastic used (Tables 13.16 and 13.17), by the wall thickness, and shape (chapter 19). Some advantages of RM are as follows:

1. Parts produced by RM are seamless.
2. Parts produced by RM are stress-free, and therefore they have excellent impact strength in large sections.
3. The lead time for the manufacture of molds is relatively short.
4. A mold can have one or more cavities.

Property	Change
Barrier properties	Increasing
Chemical resistance	Increasing
Creep resistance	Increasing
Ductility	Decreasing
ESCR	Decreasing
Hardness	Increasing
Heat deflection	Increasing
Impact strength	Decreasing
Optical properties	Decreasing
Shrinkage	Increasing
Stiffness	Increasing
Tensile strength	Increasing
Weatherability	No trend

Table 13.16 Property changes with increasing PE density (chapter 2)

Property	Change
Barrier properties	No trend
Bulk viscosity	Decreasing
Chemical resistance	Decreasing
Creep resistance	No trend
Ductility	Decreasing
Ease of flow	Increasing
ESCR	Decreasing
Flexural modulus	Decreasing
Hardness	No trend
Impact strength	Decreasing
Molecular weight	Decreasing
Stiffness	No trend
Tensile strength	Decreasing
Weatherability	Decreasing

Table 13.17 Property changes with increasing melt index (chapter 22)

5. Using two or more molds, similar or different types of products can be molded together. The usual design employs similar molds, as shown in Figure 13.13. Figure 13.14 shows a dual system with different-size molds.
6. A multilayer product can be RM (Fig. 13.15). The molding of two or more different types of plastics in a single product may be accomplished to combine their specific properties, or create a better performing or lower cost product. This process, also

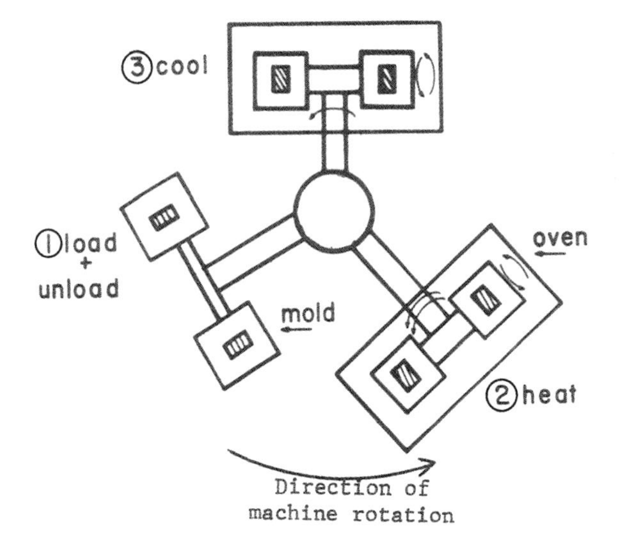

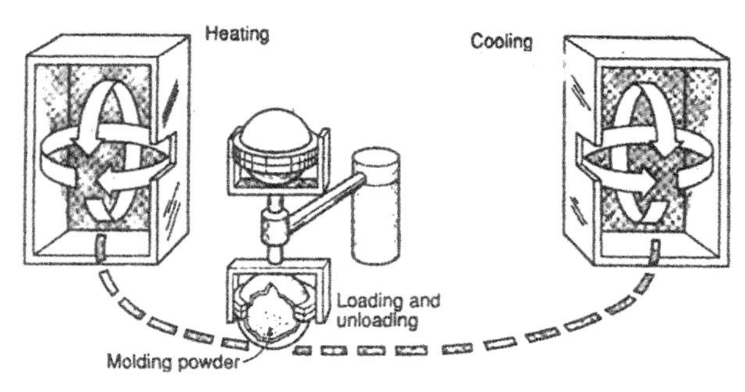

Figure 13.13 Examples of similar-mold RM machine schematics.

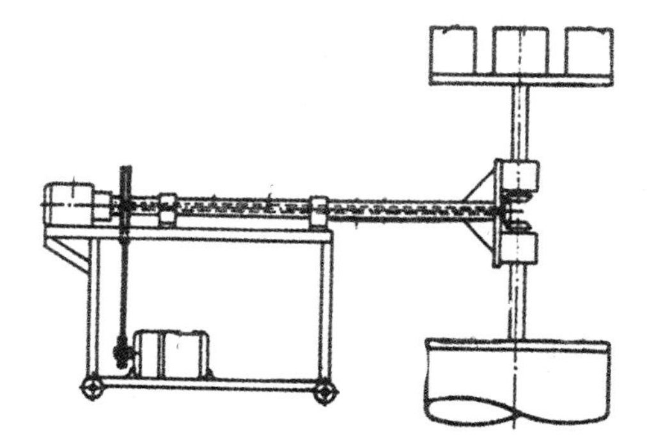

Figure 13.14 Dual system with different-sized molds.

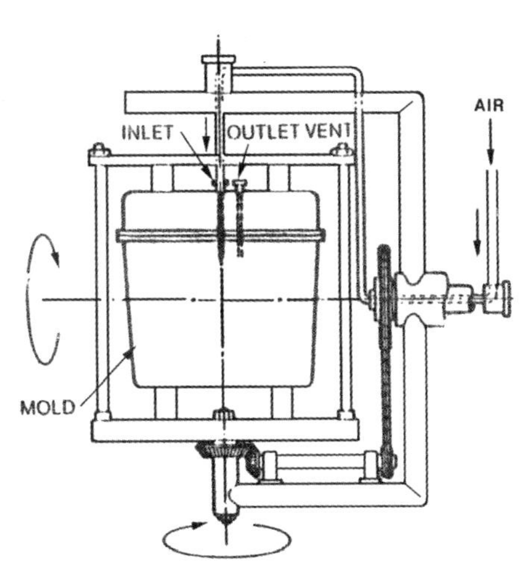

Figure 13.15 Schematic example of a multilayer RM machine.

called corotation, is similar to coextrusion or coinjection in terms of the performance of the designed product. An expensive plastic may be backed with a less costly material (recycled paper, for example), and a skin surface layer can be backed with a foamed plastic molded in one operation. The dissimilar molding powders, which may have different softening temperatures, can be molded simultaneously or separately, depending on the processing conditions and the end product's requirements. Any greater-than-normal thickness is usually used to design multilayered products, especially if a foam component is to be included. Some combinations of materials are not feasible with this method. For instance, after molding the first layer against the mold wall, the second material cannot have a higher melt temperature, which, of course, would melt the first layer, probably causing the materials to mix.

7. High-quality graphics can be molded in.
8. The wall thickness in a mold can be changed just by increasing the amount of plastic put into the mold, because the wall is produced by a coating or plating process that operates on the inside surface of the mold. However, changes in heating time would be necessary to fuse the plastic properly. Thus adjustments to a product's wall thicknesses can be made to increase rigidity, impact strength, or load-carrying capacity. A maximum thickness does exist, based on the type of material used, the material chosen to construct the mold, and the heat source. These factors all influence the rate of heat transfer through the plastic. Because in this process the plastic is deposited on the mold without pressure, the finished part is generally stress free.
9. Parts produced by RM are comparatively inexpensive, due to low mold cost. The reasons are as follows:

- Unskilled operators can operate machines.
- From none to minimum scrap loss: there is no plastic wastage in that the full charge of plastics is normally consumed in molding the products. The result is 100% plastic yield on products
- No intermediate processing is required.
- Minimum downtime to change molds.
- Low production quantity runs are economically viable.
- Compared to other processes, very large parts can be molded relatively inexpensively. Molds are not subjected to much pressure during molding, and inexpensive and thin sheet-metal molds can thus be used in many applications. Lightweight cast aluminum and electroformed or vapor-formed nickel molds, which are light in weight and low in cost, are also used.
- Mold costs are much lower in comparison to those for other processes. However, the cycle times are longer.
- Short-run, large items, and hollow shapes and products whose shapes do not lend themselves to the more popular processes such as injection molding, are usually RM.
- Metal inserts can be molded into parts.

There is one inherent overall disadvantage with RM. It is that the complete cycle for a single mold is significantly longer than it is for many other processes. However, in many cases it is possible to run multiple molds on each arm or arms to offset the effect of slower cycles. In many applications, the total cost (mold, operator salaries, etc.) is lower than that of other processes.

When compared to that of other processes, the choice of molding materials is limited. Plastic costs are relatively high due to the need of special additives and that they must be ground to a fine powder. Factors to consider include the following: (1) parting line locations, (2) part and molding machine size, (3) shot size, (4) vent location, (5) fill ports, (6) material options, (7) color matching (556), (8) tooling construction, (9) weekly volume considerations, and (10) expected design changes.

Products in this process can have deep sections and relatively sharp corners. Flat surfaces, particularly large surfaces with relatively uniform wall thickness, are difficult—if not impossible—to produce using RM. This process can be used to mold complex products that may require three or four split molds. Different finished surfaces are obtained. For example, the products' surface finish is dictated by the inside surface of the mold. This makes it easy to obtain smooth and textured surfaces on the product. Raised or depressed letters, fluting, and other decorative inscriptions may also be molded.

The inside surfaces are influenced by the type of plastic used and may be made smoother by selecting an easily flowing melt with a high melt index. Because such plastics are sometimes chemically or mechanically inferior, the better plastic may be made smoother by resorting to higher molding temperatures and longer cycle times—so long as the plastic isn't damaged. In-mold decorating methods, such as decals that are deposited on the mold surface, can become part of the finished product's surface and can be designed to provide increased structural performance. This is also true of injection molding, blow molding, and other processes.

The preferred contour for any parting line is the straightest path possible. In this way, mold construction costs can be reduced and demolding will be easier. When two products such as a container and its lid are molded together, as in blow molding, they can be separated using a removable cutter or annular wedge at the separation line. Another technique is molding the combined product oversized, to provide a resting flange, then cutting to separate the products.

Ample draft is suggested on side walls to facilitate product removal. A recommended minimum for most plastics is 1 degree (Tables 13.18 and 13.19). The lower-shrinkage plastics, such as PC and polymethyl methacrylate (PMMA), will require 1½ to 2 degrees. Undercuts are possible, but they should be kept to a minimum. Making provisions for undercuts usually requires higher mold costs because of having to use some type of action such as core pulls or splitting a mold to allow separation

Polymer	Female or Outer Draft Angle (degree)	Male or Inner Draft Angle (degree)
LLDPE	0 to 1	1 to 2
HDPE	0 to 1.5	1 to 2.5
PP	0 to 1.5	1 to 2
EVA	0 to 1	1 to 2
FPVC	0 to 1.5	1 to 3
Nylon 6	1 to 2	1.5 to 3.5
PC	1.5 to 2.5	3 to 5
PBT	1 to 2	1.5 to 3

Table 13.18 Recommended draft angles for RM plastics

Polymer	Smooth Mold (degree)		Textured (degree)	
	Female	Male	Female	Male
PE	1	2	3	6
FPVC	1.5	3	3.5	7
PC	2	4	4	8
Nylon 6	1.5	3	3.5	7
PBT	1.5	3	3.5	7

Table 13.19 Recommended draft angles for smooth and textured (0.1 mm texture depth) molds

parallel to the undercut groove. Undercuts may also require extra time for unloading molds. The more uniform the product's wall thickness becomes, the more uniform the shrinkage. However, even with very uniform products, warpage can result. The product ends are constrained by the mold corners while the centers of the flat surfaces pull away from the mold walls; this causes warpage. Table 13.20 gives examples of what can be referred to as industry standards for warpage (105).

Inside or outside corners should have large radii that are not sharp, even though sharp corners can be molded. By doing so, any potential cracking, molded-in stresses, and undesirable thickening will be prevented. A useful guide to the smallest allowable inside radius is 1/16 in (0.16 cm) with 1/4 in (0.66 cm) for optimal filling conditions (Table 13.21). The goal should be to have a radius equal to the wall thickness for easier melt flow. Although RM produces uniform wall thicknesses, corners may show greater variation than the rest of the product. A sharp inside corner tends to heat at a slower rate, causing the plastic to flow away from it, thus making it thinner. Conversely, sharp outside corners heat at a faster rate and tend to hold the plastic longer, thus building up more thickness. There are techniques to reduce or eliminate these types of problems by controlling the heat input at these corners.

It is usually difficult to produce internal or external bosses and T sections because they are not conducive to producing uniform walls. It is possible to produce interior extensions by placing a metallic screen in contact with the inner mold wall. This screen heats up, attracts plastic, becomes

Polymer	Ideal	Commercial	Precision
Polyethylene	5.0	2.0	2.0
Nylon [PA]	1.0	0.5	0.3
Polypropylene	5.0	2.0	1.0
PVC Plastisol	5.0	2.0	1.0
Polycarbonate	1.0	0.5	0.3

Table 13.20 Examples of warpage standards for RM plastics

Polymer	Inside or Female Radius (mm)			Outside or Male Radius (mm)		
	Ideal	Commercial	Minimum	Ideal	Commercial	Minimum
PE	13	6	3	6	3	1.5
FPVC	9.5	6	3	6	3	2
Nylon 6	19	9.5	4.75	13	9.5	4.75
PC	13	9.5	3	19	9.5	6

Table 13.21 Guide for inner and outer radiuses in RM dimensions

covered, and remains in place after molding. By using this method, a hollow product can be molded to have two or more separate chambers, with the screen being extended entirely across the inside of the mold. Use can also be made of a heat pipe (Fig. 13.16). This thermal pin conducts heat in these restrictive areas.

Molding a dome and cutting it after molding can form a hole. One technique that can be used for this design is to mount a fluorocarbon PTFE plug securely to the inside mold wall to prevent plastic from adhering to the mold at that location. Another method involves inserting machined brass plugs, pins, or tubing through the mold wall. During molding, the heat passes from the mold to the insert, causing plastic to form around it. Care must be taken to select inserts that will heat easily and a plastic that will not crack because stresses are created as the plastic shrinks around the insert upon cooling.

Moldable holes and inserts can complicate molding and may require extra postmolding operations. Thus the most economical designs are those that minimize the number of holes. With many plastics, both external and internal threads can be molded, but sharp V threads should be avoided because they can cause the plastic to bridge, resulting in incomplete thread fill. Rounded or modified buttress threads will allow improved thread fill.

Stiffening of solid ribs or projections is possible and in fact easy if uniform wall thickness is maintained. A narrowed rib will not fill and will leave inside stringers. It also can prevent the melt from reaching the bottom before fusing. A small, shallow, and narrow rib will fill completely but have limited strengthening effect. The correct rib design requires a wide gap to form the rib, with a generous draft so that the melt is allowed to fill uniformly, without bridging (chapter 19). A guideline is that deep ribs that are four times the wall's thickness generally require at least five times the wall's thickness between the parallel sides of the rib to prevent the plastic from bridging as it flows into the rib before fusing to the mold's wall.

MACHINES

RM machines provide different capabilities. Two types of machines are the batch machine and the carousel machine. The batch type is manually operated, going into an oven followed by the cooling

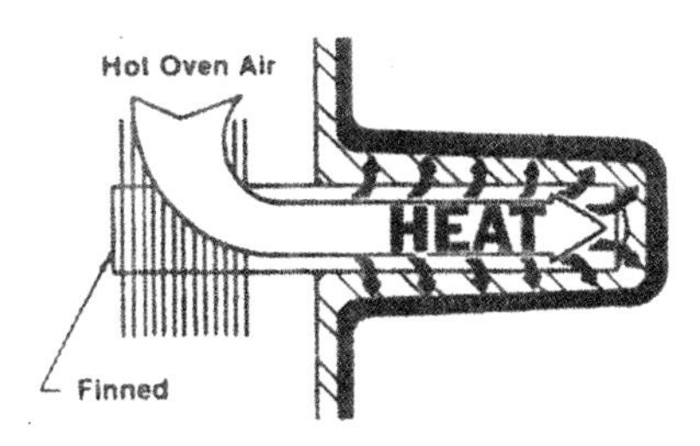

Figure 13.16 Transfer of additional heat using a heat pipe.

station. The most common is the carousel type, which hasthree stations (or four stations, if loading is considered a station): heating, cooling, and product removal, followed by reloading the plastic material (Fig. 13.17). Three cantilever arms 120 degrees apart are used on a central turret so that as one arm with a mold leaves a station, another follows into that station. All operations are automatically operating. There are also four-arm machines that can provide a second oven, cooler, or load station, depending on which is the most time consuming, so that the cycle time can be reduced.

The different designs are used to meet different processing requirements. They include double-axis rotation, carousel, shuttle, clamshell, rock-and-roll machines, and so on. These designs are similar to other molding systems in that multiple molds can be used to speed up or even simplify production.

Double-axis RM uses two platforms to hold molds. These platforms can have one or more molds. A simplified schematic view shows (1) the driving motors and variable speed transmission, (2) a hollow (outer) shaft, (3) an inner shaft, (4) mold platforms, (5) molds, and (6) the wall of a heated chamber. There are carousels that have three to six spindles (arms) for mounting molds.

Shuttle machines are principally used for RM of large products such as tanks (Fig. 13.18). A frame for holding one mold is mounted on a movable table. The table is on a tract that allows the mold and the table to move into and out of the oven. After the heating period is complete, the mold is moved into an open cooling station. A duplicate table with a mold moves into the heating oven, usually from the opposite side of the oven. As one mold is being cooled, the other mold is in the heating stage, and so on.

A very productive high-speed shuttle machine from STP Equipment (based out of Atlanta) provides advantages such as using 50% or less floor space. This single machine combines the configuration of an inline shuttle system with a rocking oven design. The linear platform, oscillating oven technology, and multiple (three or four) independent heating zones all contribute to operating efficiency.

The clamshell machines have only one arm (Fig. 13.19). The same location provides mold loading, heating, cooling, and unloading. It uses an enclosed oven that also serves as the cooling station.

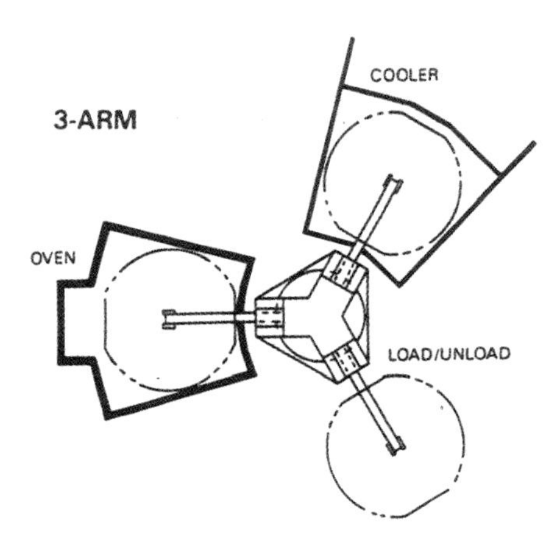

Figure 13.17 Schematic of a basic three-station RM machine.

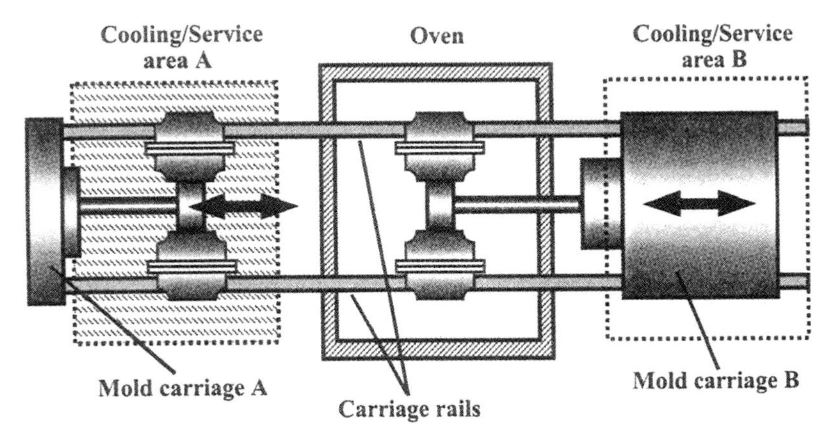

Figure 13.18 Example of a shuttle machine.

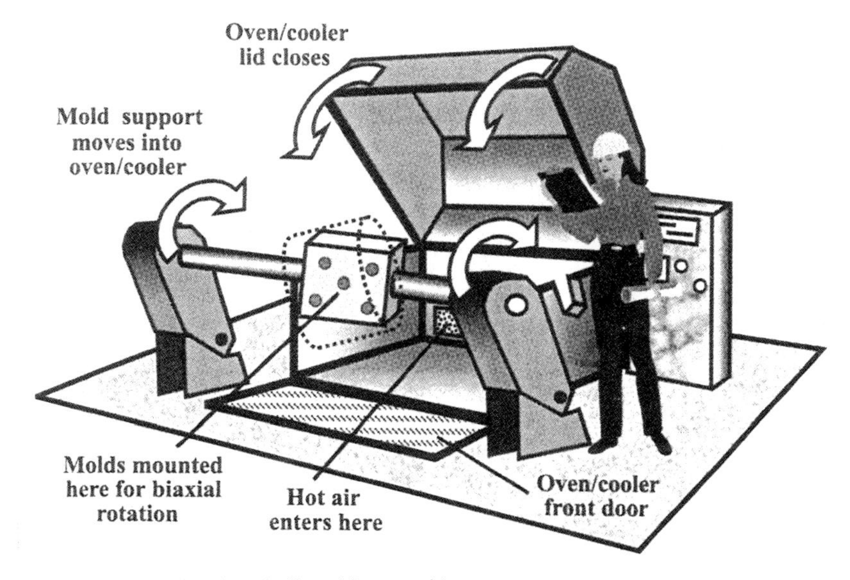

Figure 13.19 Example of a clamshell molding machine.

The rock-and-roll technique is popular for molding long products such as canoes, surfboards, and kayaks. It rotates on one axis and tilts to provide action in the other axial direction (Fig. 13.20). Most of the motion is in the long direction of the product, which relates to the main rotating motion. These open-flame (and slush molding) machines are the oldest types of equipment used. They basically rotate only on one axis, fabricating open-ended containers, rubber raingear, shoes, and other products. The material moves back and forth (slushing action) during the heating cycle.

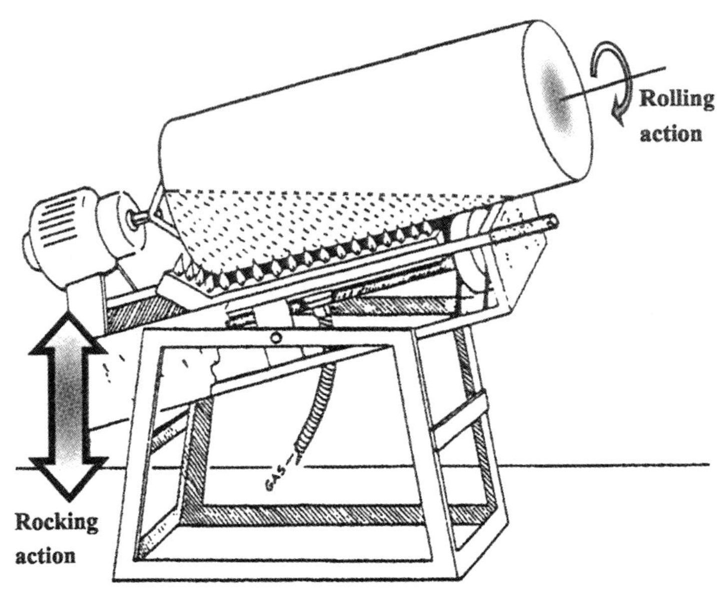

Figure 13.20 Example of a rock-and-roll molding machine.

RM machines are becoming larger and more electronically sophisticated. Larger machines generally require more sophisticated electronics to provide increasingly precise control of the molding process.

Microprocessors that indicate processing set points and conditions improve the precision of the product. The molding cycle data for the oven and cooling chambers can be stored for different products and recalled when needed. Cycle time, oven temperature, major and minor axis speeds, and fan and water-spray times are typical of functions under complete computer control. Systems can involve an infrared thermometer to monitor the mold's outer temperature. This permits optimizing the process by controlling the machine cycle based on temperature and time, rather than on time alone.

Arms can be equipped with inner lines capable of bringing hot air, nitrogen, or vacuum into a mold during the heating cycle. The same lines can bring pressurized air into the mold during the cooling cycle. There are also special machines that deviate from the use of a traditional oven. This type of machine uses conductive composite molds rotating biaxially on a frame to form products without using an oven or an open flame.

There is a way to cut labor costs and eliminate over- or undercharging molds. There are systems that precisely dispense up to four material components into a mold. The plastic is dispensed up to 6 lb/s with a reported accuracy of ±0.02 lb. The system works as the feeder automatically reads a barcode. The barcode is matched to a dispensing recipe and then the required amounts of color and plastic powder are dispensed. The system is capable of holding more than 65000 recipes.

MOLD

Most molds are thin shell-like structures made from metals (Table 13.22; chapter 17). Molds made from other materials (plaster casting [Table 13.23], reinforced plastics [chapter 15], wood, etc.) are used for special applications. Lightweight cast aluminum and clectroformed or vaporformed nickel molds, which are light in weight and low in cost, can be used. Aluminum molds are most frequently used especially, for small to medium-sized products. Aluminum has better heat transfer than steel and is very cost-effective when several molds for the same product are required (Table 13.24). The aluminum mold with good framing is frequently a good choice for structural products, as it can easily be constructed to allow contoured shapes and blended radiuses with the use of a wood pattern.

Sheet-metal molds are normally used for larger products (Table 13.25). They are easy to produce since sections can be welded together. Since the molds are not subjected to pressure during

Material	Density, ρ kg/m^3 (lb/ft^3)	Thermal Conductivity, K, W/m K (Btu/ft h F)	Specific Heat Capacity, C_p J/kg K (Btu/lb F)	Elastic Modulus, E GN/m^2 (Mlb/in^2)	Coefficient of Linear Thermal Expansion, α_T 10^{-6} K^{-1}
Aluminum	2800 (175)	147 (153)	917 (0.4)	70 (10.2)	22.5
Nickel	8830 (551)	21.7 (22.6)	419 (0.18)	179 (26)	14.1
Carbon steel	7860 (491)	51.9 (54)	486 (0.21)	206 (29.8)	12.2
Stainless steel	7910 (494)	14.5 (15.1)	490 (0.21)	201 (29.2)	16.3

Table 13.22 Properties of mold materials

Commercial Name	Source	Water Ratio (pph)	Setting Time (mins)	Dry Compressive Strength MN/m^2 (lb$_f$/in^2)
Pattern shop Hydrocal A-11	U.S. Gypsum	54–56	20–25	22.1 (3,200)
Industrial White Hydrocal	U.S. Gypsum	40–43	20–30	38 (5,500)
Ultracal 30	U.S. Gypsum	35–38	25–35	50.3 (7,300)
Densite K5	Georgia Pacific	27–34	15–20	65.5 (9,500)
Super X Hydro–Stone	U.S. Gypsum	21–23	17–20	96.5 (14,000)

Table 13.23 Plaster casting materials

Polymer	Oven Temperature (°C)	Thickness (mm)	Exit Temperature (°C)	Time (min)
HDPE	300	2	210	13
HDPE	300	4	210	23
HDPE	300	6	210	32
HDPE	300	8	205	43
HDPE	300	10	210	56
MDPE	275	6	210	22
PP	325(?)	3	240	18
PC	375(?)	3	265	22
PVC	200(?)	5	133	23
ABS	350(?)	3	300	17
ETFE	325	4.5	290	26
Hytrel	300(?)	3	220	13.5
Nylon 6	325(?)	3	230	16
XLPE	260	3	180	13.5
PFA	330	3	300	33

Table 13.24 Heating cycle times for aluminum molds

Gage	Thickness mm (inch)	Weight kg/m^2 (lb/ft^2)
10	3.57 (0.1406)	27.46 (5.625)
12	2.78 (0.1094)	21.36 (4.375)
14	1.98 (0.0781)	15.26 (3.125)
16	1.588 (0.0625)	12.21 (2.5)
18	1.27 (0.0500)	9.765 (2.0)
20	0.952 (0.0375)	7.324 (1.5)
22	0.794 (0.0312)	6.1 (1.25)

Table 13.25 Steel sheet-metal gauge

molding, they are not built to take the high loads required in molds for injection, compression, and other pressure-operating molds.

Two-part molds are usually used, but molds in three or more parts are sometimes required to remove the finished products. Molds can be as simple as a sphere, and molds can be complex with undercuts, ribs, or tapers or all three. Design considerations include heat transfer, mounting techniques, parting lines, clamping mechanisms, mold release (559), venting, and material stability in storage and during the RM process.

The preferred contour for any parting line is the straightest path possible. By this means, mold construction costs can be reduced and demolding will be the easiest means possible. When two partner products, such as a container and its lid, are to be molded together, as in blow molding, they may be separated after the molding by employing a removable cutter or annular wedge at the parting line. Another technique is to mold it oversize to provide a resting flange, then cutting it to separate the products.

Ample draft is suggested on sidewalls, to facilitate product removal. A recommended minimum for most plastics is 1 degree. The lower-shrinkage plastics like PC and PMMA will require 1½ to 2 degrees. Undercuts are possible, but they should be kept to a minimum. Making provisions for undercuts usually requires higher mold costs, because of having to use some type of action such as core pulls or splitting a mold to allow separation parallel to the undercut groove. Undercuts may also require extra time for unloading molds.

Automation exists in molds. There are automated mold opening and closing units. These boltless and clampless systems position the wheel-shaped spider horizontally under a hydraulic arm. This descends and actuates an unlocking cylinder that holds two or more molds shut. Springs then push the upper mold halves away from the lower halves and the arm lifts them free so that the products can be demolded. After the molds are recharged, the hydraulic arm descends, pushing the molds shut, and then actuates the locking device to hold them closed after the arm withdraws. The spider then indexes the next group of molds into place. This method is used to save time, labor, and mold damage.

Mold release agents are usually required because the plastic melt may adhere to the surface of the mold cavity, particularly if the cavity has a very complex shape with contours, ribs, and other features. Many molds must have very little or no draft, so they require a mold-release agent. There are mold-release agents that can be baked or applied to the cavity by wiping. By coating with fluorocarbon, the need for mold release could be eliminated. With conventional RM, after the initial mold-release agent is applied, several hundred products can usually be molded before the mold cavity is stripped down and another baked-on coating is applied. During this time, some touch-up of the mold may be required.

It may be necessary to sandblast the mold cavity to remove any buildup of release agent before reapplying the release agent. With the types of plastic that are generally used, melt flow of the plastic during processing will produce a smooth product surface; sandblasting does not affect the smoothness of the surface.

Like other molding operations, a textured cavity can provide a textured product surface. Most texturing of cavities is by chemical etching, so it is important to use the appropriate mold material to create a particular texture. An effective release is needed at the parting line to aid in demolding.

RM venting molds are often used to maintain atmospheric pressure inside the closed mold during the entire molding cycle. Rotational molds may require a venting system to remove the gas that develops during the heat cycle. The vent will reduce or prevent mold distortion and lower the clamping pressure needed to keep the mold closed. It will prevent blowouts caused by pressure and permit use of thinner molds. The vent can be a thin-walled tube with an internal liner of PTFE.

The opening where the vent enters the mold is located where it will not harm the performance or appearance of the molded product. The vent is usually filled with glass wool to keep the material charge from entering the vent during rotation and particularly during the heating cycle. The outside end of the vent has to be protected so that no water will enter during cooling. A simple method is to put a cap on it; the cap can be automatically controlled to close just before the melt enters the cooling chamber.

DESIGN

Of the three competing processes—single-sided thermoforming, blow molding, and RM—only RM has the potential to yield uniform wall thickness for even the most complex product. When considering mechanical design requirements, review Table 13.26.

The design of rotationally molded products requires a working relationship between aesthetics and performance. RM offers the designer a unique way of manufacturing "bulky" articles from simple balls to complex, near-parallel walled structures. Since very little pressure and shear are applied during processing, products are essentially stress free. And as noted earlier, the way in which powder is distributed and coalesced on the mold surface yields an inherently near-uniform wall thickness.

There are certain guidelines that the designer of RM products should understand. Even though RM yields inherently uniform walls when compared with thermoforming and blow molding, RM is a single-surface process similar to thermoforming and blow molding. As a result, wall-thickness tolerance is never as good as two-surface processes such as extrusion and particularly injection molding. For generic, run-of-the-mill products such as tanks and outdoor toys, RM wall-thickness tolerance is ±20%. For certain tight-tolerance products such as medical face masks and optical parts, a tolerance of ±10% can be specified, albeit with a greater percentage of rejects. As a result of this wide tolerance, in RM, as well as in blow molding and thermoforming, it is common to specify minimum wall thickness rather than nominal wall thickness.

The primary objective in any product design is to make the product capable of withstanding expected loads with appropriate safety factors, but without adding so much plastic that the product is no longer economically competitive. Table 13.27 shows approximate wall-thickness ranges for many RM products.

Instead of specifying a nominal wall thickness of, for example, 6 mm, as is common with injection molding, where the tolerance may be ±0.2 mm, the RM minimum wall thickness would be 5.8 mm with a tolerance of 0 mm to 2.3 mm. If a nominal wall thickness must be specified for this rotationally molded part, it would be 7 mm ±1.2 mm.

Wall-thickness uniformity in the final product is the result of the early processing step of tackifying. This stage is an averaging step. Once the powder begins adhering to the mold surface, slip flow disappears.

As reviewed in Chapter 19, plastic melt during cooling results in plastic shrinkage. Examples of RM plastic shrinkages are shown in Table 13.28.

- Polycarbonate and nylon powder must be kept very dry prior to molding, to prevent moisture pick-up. Moisture will degrade the polymer, resulting in lowered physical properties, particularly impact. Moisture will also lead to the formation of microbubbles, which act as stress concentrators. The presence of bubbles may also lead to reduced impact strength.
- Solid ribs cannot be successfully rotationally molded. Hollow ribs, where the rib width-to-depth ratio is greater than one, are recommended.
- Shallow undercuts are possible with polyethylene and polypropylene. Deep undercuts are possible with PVC plastisol. Undercuts are not used when molding stiffer polymers such as polycarbonate.
- Care must be taken when pulling a warm polypropylene or nylon part from the mold, since the polymer may not be fully crystallized and any distortion may become permanent.
- When determining final part price-performance ratio, thinner part walls mean shorter molding cycle times and lower material costs. However, stiffness reduces in proportion to the part wall thickness to a power of three.
- Flat-panel warpage is minimized through part design. Crowns, radial ribs, domes, stepped surfaces, and corrugations will act to minimize warpage.
- If warpage is severe, the cooling rate during molding must be reduced. If warpage continues to be severe, mold pressurization may be required.
- Rotational molding is used to make parts with parallel or near-parallel walls. The distance between the walls must be sufficient to allow for powder flow and to minimize bridging. The distance between walls should be at least three times the desired wall thickness. Five times is recommended.
- If the part is bridged in a given region, it will take longer to cool in that region. The result will be generation of internal voids and differential shrinkage, which may lead to part distortion and localized sink marks. For the most part, rotational molding yields stress-free parts. However, in bridged areas, local stresses may be quite high and may lead to local part failure in fatigue or flexure.

Table 13.26 RM mechanical design aspects

- If the depth of the outer mold cavity is greater than the width across the cavity, heat transfer to the bottom of the cavity may be restricted. The result will be that the wall thickness on the inside of the double wall may become very thin, especially at the very bottom of the wall. Stationary baffles on the mold surface are effective for cavities with depth-to-width ratios less than about 0.5. Forced air venturis are currently recommended for deeper cavities.

- Insulation pads are applied to a local area to minimize thickness in that area. Regions where little or no plastic is desired would include areas to be trimmed on the final part. If the part needs to have a thicker wall in a given area, the mold wall is made thinner or the mold is made of a higher thermal conductivity metal in that area.

- Small-radius inside mold corners typically take longer to heat and cool and therefore part walls can be thinner in corners than in adjacent sidewalls. Generous radii mitigate this problem. Small-radius outside corners tend to heat and cool more rapidly and therefore part walls can be thicker in corners than in adjacent sidewalls. Again generous radii mitigate this problem.

- Structural strength is obtained primarily through addition of stiffening elements such as chamfered or large-radiused corners, hollow gussets, hollow ribs, and round or rectangular kiss-offs (or almost-kiss-offs). For hollow double-wall parts such as decks and doors, it is desired to have indentations such as ribs and kiss-offs molded in both surfaces. This aids in energy distribution to and minimizes thinning at the bottoms of the ribs and kiss-offs. The widths of the openings of the indentations must be increased if the design requires that one surface be indentation-free. Addition of fillers or reinforcing fibers as stiffening agents is not recommended in rotational molding.

- Rim stiffening is achieved by adding ribs just below the rim, or by flanging the rim with either a flat flange or a U-shaped flange. A metal reinforcing element, such as a hollow conduit, can be placed in the mold prior to powder filling. This allows the reinforcing element to be an integral part of the structure. The designer must remember that plastics have about 10 times the thermal expansion of metals and that the metal must be affixed so that it does not create concentrated stresses on the plastic part during heating and cooling.

- As detailed below, there are many reasons to have large-radiused corners. Outside corners on parts tend to shrink away from the mold wall and so have low residual stresses. Inside corners on parts tend to shrink onto the mold wall and so have greater residual stresses than neighboring walls.

Table 13.26 RM mechanical design aspects *(continued)*

- Deep undercuts are formed around removable inserts or core pins. These are made either of a high thermal conductivity metal such as aluminum for a steel mold or copper-beryllium for an aluminum mold, or are hollowed out.
- Rotationally molded parts usually are formed in female molds at atmospheric pressure, with shrinkage allowing the part to pull away from the mold. This allows parts to be molded with no draft angle and thus vertical sides.
- Although rotational molding uses no pressure, the polymer against the mold wall is molten. As a result, it is possible to transfer quite fine texture from the mold wall to the finished part. Competitive processes such as thermoforming and blow molding require differential pressures of 3 to 10 atmospheres to achieve similar results.
- Deep undercuts, including complex internal threads, are possible through proper mold design.[5]
- Inwardly projecting holes can be molded in using core pins. If the pin is long enough or if it is solid, the polymer will not cover the pin end. If the pin is short, hollowed out, or is a thermal pin where heat is rapidly conducted down the pin length from the oven air, the hole will be blind. Large diameter outwardly projecting holes are possible, as long as the diameter-to-length is less than one and the diameter-to-wall thickness is greater than about five. Outwardly projecting holes are molded closed and are opened with mechanical means such as saws or routers. Holes should be spaced about five wall thicknesses from each other.
- Detents molded into the part wall provide locators for drills and hole saws.
- Both internal and external threads can be rotationally molded into parts. The recommended thread design is the "modified buttress thread profile" or Acme thread. For fine-pitched, sharp threads, or for small-diameter threads, an injection-molded thread assembly is placed in the mold prior to powder filling. The powder melts and fuses the assembly to the part body.
- In many instances, the rotationally molded part must be assembled to other parts using metallic screws or fasteners. Metal inserts have been developed especially for rotational molding. These inserts, usually of a high thermal conductivity metal, are placed in the mold prior to powder filling. Powder melts and fuses the insert to the part body. As the polymer shrinks, it is compressed around the insert, holding it in place. However, the metal prevents the polymer from shrinking fully. As a result, residual stresses are imparted in the insert region. These stresses can be a source of part failure during use. To minimize webbing and undue stress concentration, metal inserts should be three to five wall thicknesses away from corners.

Table 13.26 RM mechanical design aspects *(continued)*

Polymer	Minimum Wall Thickness (mm)	Typical Wall Thickness Range (mm)	Maximum Wall Thickness (mm)
LLDPE	0.5	1.5 – 25	75
HDPE	0.75	1.5 – 25	50
FPVC	0.2	1.5 – 10	10
Nylon 6	1.5	2.5 – 20	40
PC	1.25	1.5 – 10	10
EVA	0.5	1.5 – 20	20
PP	0.5	1.5 – 25	25

Table 13.27 Wall-thickness range for RM plastics

Polymer	Shrinkage Range (%)	Recommended (%)
LDPE	1.6 – 3.0	3.0
HDPE	3.0 – 3.5	3.5
PP	1.5 – 2.2	2.2
FPVC	0.8 – 2.5	1.5
PC	0.6 – 0.8	0.8
CAB	0.2 – 0.5	0.5
Nylon 6	1.5 – 3.0	3.0

Table 13.28 Guide to linear shrinkage values for RM plastics

A summation to the design details follows: (1) keep high-stress areas away from parting lines; (2) design for very high coefficients of thermal expansion; (3) avoid flat surfaces with straight lines; (4) change deep pocket areas for better heat flow and thicker walls; (5) keep tolerances very high, typically larger than handbooks accept; (6) expect that occasionally the amount of powder required for the material thickness chosen may not fit into the design shape, thus a dropbox may be required; (7) keep unit loads at component mounts below 100 psi; (8) keep parting lines above or below natural eye level; (9) recognize that heat-deflection temperatures are typically low, about 150°F (chapter 5); (10) model or draw the part with all the drafts shown; (11) keep inserts away from sidewalls for good material fill; and (12) conduct a design review again before the end of the design phase, as details have probably changed since the planning stage, and complete the design checklist at this point.

The Association of Rotational Molders (ARM) has published a design manual for RM. ARM represents the rotomolding industry internationally. It includes molders, plastic and equipment manufacturers, design organizations, and professional consultants. ARM is also the major information-disseminating organization in the RM industry. It has compiled a comprehensive library with the industry's design manual. Another important organization on this subject is the Rotational Molding Development Center (RMDC) at the University of Akron. It was founded in 1986 to provide for the industry's future research needs.

PRODUCTION AND COST

Its production rates, compared to those of other processes, RM can be low. The total cost of equipment and the production time for moderate-sized and especially large products are also low.

Molds can be of any shape and can include corrugated or rib constructions to increase their stability and stiffness (large, flat walls can be difficult if not impossible). As reviewed, the thickness of their walls is limited to allow heat penetration.

This process is a particularly cost-effective for molding small to long production runs of small to large products and is especially cost-effective for very large products. The molds are not subjected to pressure during molding so they can be made relatively inexpensively out of thin sheet metal. The molds may also be made from lightweight cast aluminum and electroformed nickel, both of which are light in weight and low in cost. Large rotational machines can be built economically because they use inexpensive gas-fired or hot-air ovens with the lightweight mold-rotating equipment.

Rotational manufacturing plants are located worldwide. The latest information on US plants regarding number of machines, output rates, markets served, number of employees, personnel involved, yearly sales, and addresses are provided by *Plastics News*. This listing, as well as others that *Plastics News* publishes about other processes, materials, and markets, is updated about every year. *Plastics News* reported that the largest custom rotomolder was Centro Inc. in North Liberty, Iowa. They were the fourth-largest firm overall in *Plastics News*'s ranking of North American rotomolders, with $59 million in sales, 625 employees, and 27 machines in six factories. The company was founded in 1970.